W9-CLG-901

Point Mapping Stability

by

JACQUES BERNUSSOU

Docteur ès-Sciences

With the collaboration of

J. L. ABATUT, J. P. BABARY, A. GIRAUD, LIU HSU, and F. ROUBELLAT

Laboratoire d'Automatique et d'Analyse des Systemes

du

Centre National de la Recherche Scientifique, Toulouse

PERGAMON PRESS

OXFORD · NEW YORK · TORONTO · SYDNEY · PARIS · FRANKFURT

U.K.	Pergamon Press Ltd., Headington Hill Hall, Oxford OX3 0BW, England
U.S.A.	Pergamon Press Inc., Maxwell House, Fairview Park, Elmsford, New York 10523, U.S.A.
CANADA	Pergamon of Canada Ltd., 75 The East Mall, Toronto, Ontario, Canada
AUSTRALIA	Pergamon Press (Aust.) Pty. Ltd., 19a Boundary Street, Rushcutters Bay, N.S.W. 2011, Australia
FRANCE	Pergamon Press SARL, 24 rue des Ecoles, 75240 Paris, Cedex 05, France
WEST GERMANY	Pergamon Press GmbH, 6242 Kronberg-Taunus, Pferdstrasse 1, West Germany

First edition 1977

Library of Congress Catalog Card No. 77-78803

In order to make this volume available as economically and rapidly as possible the author's typescript has been reproduced in its original form. This method unfortunately has its typographical limitations but it is hoped that they in no way distract the reader.

Printed in Great Britain by William Clowes & Sons, Limited London, Beccles and Colchester

ISBN 0 08 021283 2

CONTENTS

FOREWORD

Research in Non-Linear Mechanics, in particular on systems described by recurrence equations (known as pointwise transformations or as difference equations) point mapping was started in the Laboratoire d'Automatique et d'Analyse des Systèmes (L.A.A.S.) of the Centre National de la Recherche Scientifique in Toulouse in 1963 at the instigation of Professor C. MIRA of the Institut National des Sciences Appliquées. The authors of the present work have all at different times been members of the research team which was set up at that period, and have each presented theses in this area. This monograph is the result of the efforts of one of the research workers, J. BERNUSSOU, to bring together the most significant results to emerge from these theses.

The International Colloquium of the C.N.R.S. on "Transformations ponctuelles et applications", held at the L.A.A.S. in 1973 attracted experts from a wide variety of disciplines (astronomy, particle physics, automatic control, applied mathematics, etc.).

I feel therefore that this monograph should be of interest both to engineers and established research workersas well as to postgraduate students in these domains.

The work described here is situated not so much in Mathematics as in the area between Applied Mathematics and Engineering Science ; the reader will thus find, in addition to the purely theoretical sections, examples which bring out the possible applications of what may be called the Associated Pointwise Transformation Method.

Jean LAGASSE

PREFACE

The progress of the mathematical sciences corresponds in some ways to the discovery and development of unknown countries in the past : some explorers aimed at a new target, showed the ways and the directions - but in most cases the actual conquest was performed by the ones who followed the first pioneers, extended the ways, developed better methods of work, created new cross connections, and, took possession of the whole region, gradually.

The work being done at the Toulouse "Laboratoire d'Automatique et d'Analyse des Systèmes" is now presented in book form. Its results might be compared with the act of taking possession of the new province of mathematics, into which H. Poincaré made his first advances some eighty years ago . In his investigations of dynamical system in the Euclidean R^n described by autonomous differential equations he considered the points where the trajectories cross a certain hyperplane and had the idea of regarding the sequence of these points as a mapping of the hyperplane onto itself. This "point transformation" which can be described by means of a difference equation then became the actual object of research. Although the investigations started with this rather simple mathematical idea they have led to many more complex questions, e.g. : "To what extent do qualitative properties of the 'discrete' trajectories of the point transformation correspond to the properties of the 'continuous' trajectories of the original system ?" - "What results can be deduced from the type of the singularities of the difference equation ? " - "Is it possible to obtain quantitative results for the differential equation by applying quantitative methods to the point transformations ?" - "Is it possible to reverse the relationship, i.e. to construct a differential equation associated with a given point transformation ?" - This has happened in many branches of mathematics : a solved problem has led to new questions, and auxiliary concepts arouse interest of their own and themselves become objects of research.

A universal theory of point transformations could obtain only a few general theorems and would not help at all. But one cannot even anticipate finding general methods for dealing with the problems in question since they are mostly non linear, and the common feature of non linear problems is not a property but the lack of a certain property, namely, the linearity of the principle of superposition. Therefore one has to consider in each special case the method which might be applied to obtain further results. And in many instances one will probably find that none of the known methods is useful at all.

The team working in Toulouse has been successful in dealing with these questions. Especially, they extended known methods and developed new techniques. In addition, they applied their investigations to special problems in Physics and Engineering and used numerical calculations to check their theoretical results.

For the specialist it will be most valuable to find many important results- usually scattered over numerous publications - in one monograph. I do hope that it will stimulate non-specialists to work in this new branch of non linear analysis.

Wolfgang HAHN
Technical University Graz

ACKNOWLEDGEMENTS

We are deeply grateful to Professor J. LAGASSE who has been until the present time Director of the Laboratoire d'Automatique et d'Analyse des Systèmes du Centre National de la Recherche Scientifique for giving his support and encouragement both to our research and to the publication of this monograph. We should also like to express our sincere thanks to Professor C. MIRA of the Institut National des Sciences Appliquées de Toulouse for supervising this work over several years. In addition, we are most grateful to Professor I. GUMOWSKI, currently visiting the Université Paul Sabatier, Toulouse, who defined certain of the problems dealt with in this book. Finally we should like to thank Professor W. HAHN for the interest he has so kindly shown in this monograph.

CHAPTER I
FUNDAMENTALS ON POINT MAPPINGS

This relatively short chapter consists of a brief survey of the fundamental properties of recurrent equations, or point mappings and introduces the main questions that will be discussed in the following chapters.

Dynamical systems are generally described using differential equations

$$\frac{dX}{dt} = F(X) \tag{1.1}$$

in the autonomous case where the independent time variable does not appear explicitly on the right hand side of the equation

$$\frac{dX}{dt} = F(X,t) \tag{1.2}$$

in the non autonomous case.

If X is an m dimensional vector, and the usual existence and uniqueness conditions are satisfied, then the solutions of (1.1) and (1.2) can be represented in an m dimensional space as trajectories i.e. the locus of points whose coordinates are the components of X(t) in the space.
The notion of dynamical systems can be extended to cover systems in which the independent variable is in discrete time. Systems of this kind are described using recurrent equations

$$X_{n+1} = G(X_n) \tag{1.3}$$

in the autonomous case

$$X_{n+1} = G(X_n,n) \tag{1.4}$$

in the non autonomous case.

By means of successive iterations of equations (1.3) and (1.4) from an initial condition (X_0,n_0), a series of points is obtained giving a discrete trajectory in the m dimensional space (assuming the existence and uniqueness conditions to be satisfied). This is the origin of the term point mapping : using (1.3) or (1.4) the point $M_n(X_n)$ is mapped on to the point $M_{n+1}(X_{n+1})$.

I.1. Some of the Main Motivations for Studying Point Mappings

There is a large number of systems whose behaviour can be studied using associated point mapping. The principal reasons for this are as follows : firstly, recurrent equations are the natural mathematical model for describing certain physical systems among which are to be found systems with discrete information : sampled data systems, pulse modulated systems (pulse width modulation, pulse frequency modulation...). The use of numerical machines in process control is a further example of such a system. In addition, in scientific calculus, the numerical treatment of differential equations (numerical integration) and the related problem of sensitivity to discretisation has led to the study of recurrent equations obtained by discretisation to see how far their behaviour is a consequence of disturbances caused by discretising.

The second reason is in fact the one which originally motivated the development of the study of point mappings. By introducing the notion of surface of section, Poincaré (1) showed that the study of an m dimensional dynamical system (autonomous case) is reduced to the study of an (m-1) dimensional autonomous mapping of the surface on to itself. The first obvious advantage of this approach is that it reduces the order of the equations. It is widely used in celestial mechanics (chapter 4), since this is an area in which the reduction of dimensionality is more important.

In the case, for example, of Hamiltonian autonomous dynamical systems with two degrees of freedom, use of the first integral of motion constituted by the Hamiltonian means that the study is reduced to a second order mapping of a two dimensional surface of section on to itself. In the work of Andronov et al (2) several examples are given of second order piecewise linear systems whose dynamics is studied by the point transformation method (first order point mapping). The subject matter of chapter 4 (periodic equations) can also be considered as an application of the point mapping approach to systems not directly described by means of recurrent equations. Here also the reduction in the order of equation is significant since an autonomous recurrent equation can be associated with a non autonomous differential equation, this reduction being accounted for by the disappearance of the **independent** time variable.

The next section will be devoted to some general definitions which will clarify some of the concepts used in the rest of the work.

I.2. Some Definitions

What follows will apply to uniform, autonomous, continuous recurrent equations. Each of these terms refers to the real valued functions of the real variables which appear in the recurrent equations :

$$X_{n+1} = F(X_n), \quad X_n \in R^m \tag{1.5}$$

(1.5) defines the point mapping T :

$$M_n\ (X_n) \xrightarrow{T} M_{n+1}\ (X_{n+1})$$

M_{n+1} is called the consequent of the point M_n, M_n the antecedent of M_{n+1}. The antecedent is defined by application of the inverse mapping T^{-1}, it should be noted that this antecedent is not necessarily unique. The series of points obtained by iteration of T is called a discrete trajectory.

A double point (or fixed point) is a point invariant by T, that is a point whose coordinates satisfy :

$$X_{n+1} = X_n = F\ (X_n) \tag{1.6}$$

A cycle of order k is a set of k points that are double points for T to the power k (T^k) and not invariant by T^e, with e integer < k.

$$X_{n+k} = X_n \quad ; \quad X_{n+e} \neq X_n \quad \text{for} \quad e < k \tag{1.7}$$

It may happen that equations (1.6) or (1.7) are satisfied by the coordinates of an infinity of points defining a curve of double points or cycles. The double points, the cycles of order k are also called singular points or singularities of the mapping (1.5).

I.2.1. Characteristic Equation - Multipliers - Stability

Let $M_0(a)$ be a double point for (1.5). Provided F is differentiable in the neighbourhood of X=a, (1.5) is written

$$X_{n+1} = a + A\,(X_n - a) + R^2(X_n - a) \tag{1.8}$$

where A is the matrix (m x m) of partial derivatives of F(X), calculated at M_0, $R^2(.)$ being the remaining terms of degree equal at least to 2. The characteristic equation of the double point is then : det[A-SE] = 0, where E is the identity matrix. The multipliers of the mapping (1.5) at M_0 are the roots of the characteristic equation, i.e., the eigenvalues of the Jacobian matrix of (1.5) at X=a.

It is well known that the double point M_0 is :
- asymptotically stable if and only if all the multipliers are smaller than one in absolute value,
- unstable if the absolute value of at least one of the multipliers is greater than one.

The remaining cases in which the absolute values of one or more multipliers is equal to one (the other being less than one) are called critical cases in the sense of Lyapunov. These cases are delt with in chapter 2. Clearly, these results can easily be extended to the m order cycles by simply considering the m th power of the mapping ; T^m instead of T.

1.2.2. Classification of the Singularities

For a second order point mapping

$$x_{n+1} = f(x_n, y_n) \quad ; \quad y_{n+1} = g(x_n, y_n) \tag{1.9}$$

excluding critical cases, the singular points have been classified in the same way as for differential equations, according to the behaviour of the trajectories in their neighbourhood. Let S_1, S_2 be the two multipliers :

S_1, S_2 complex = $r\exp[\pm j\Phi]$, the singularity is a focus, stable (unstable) if $r<1$ $(r>1)$

S_1, S_2 real: - both absolute values less (greater) than one, the singularity is a stable (unstable) node
- $|S_1|<1<|S_2|$, the singularity is a saddle.

In the case of distinct real multipliers, two principal axes can be defined at the double points (or cycles) which are parallel to the eigenvectors of the matrix A, whose directions are given by

$$P_1 = \frac{S_1 - a}{b} = \frac{c}{S_1 - d} \quad \text{and} \quad P_2 = \frac{S_2 - a}{b} = \frac{c}{S_2 - d}$$

$$a = \frac{\partial f}{\partial x}\Big|_{M_0} \; ; \; b = \frac{\partial f}{\partial y}\Big|_{M_0} \; ; \; c = \frac{\partial g}{\partial x}\Big|_{M_0} \; ; \; d = \frac{\partial g}{\partial y}\Big|_{M_0}$$; M_0 being the double point

Clearly, the principal axes are those in which the mapping is written at M_0 in canonical form (diagonal in the case of distinct multipliers)

$$u_{n+1} = S_1 u_n + U(u_n, v_n)$$
$$v_{n+1} = S_2 v_n + V(u_n, v_n) \tag{1.10}$$

In the neighbourhood of the double point there is one or an infinity of discrete trajectories "tangential" to the principal axes. A more detailed classification of the singularities of a mapping can be made according to the sign of the multipliers. This represents a divergence from the type of classification used for continuous differential equations. Conventionally when both S_1 and S_2 are positive the node or saddle is said to be of type 1, of type 2 when S_1 or S_2 is negative, of type 3 when both are negative. This distinction is brought out in Fig. 1.1 which shows the location of the successive iterations with respect to the principal axes.

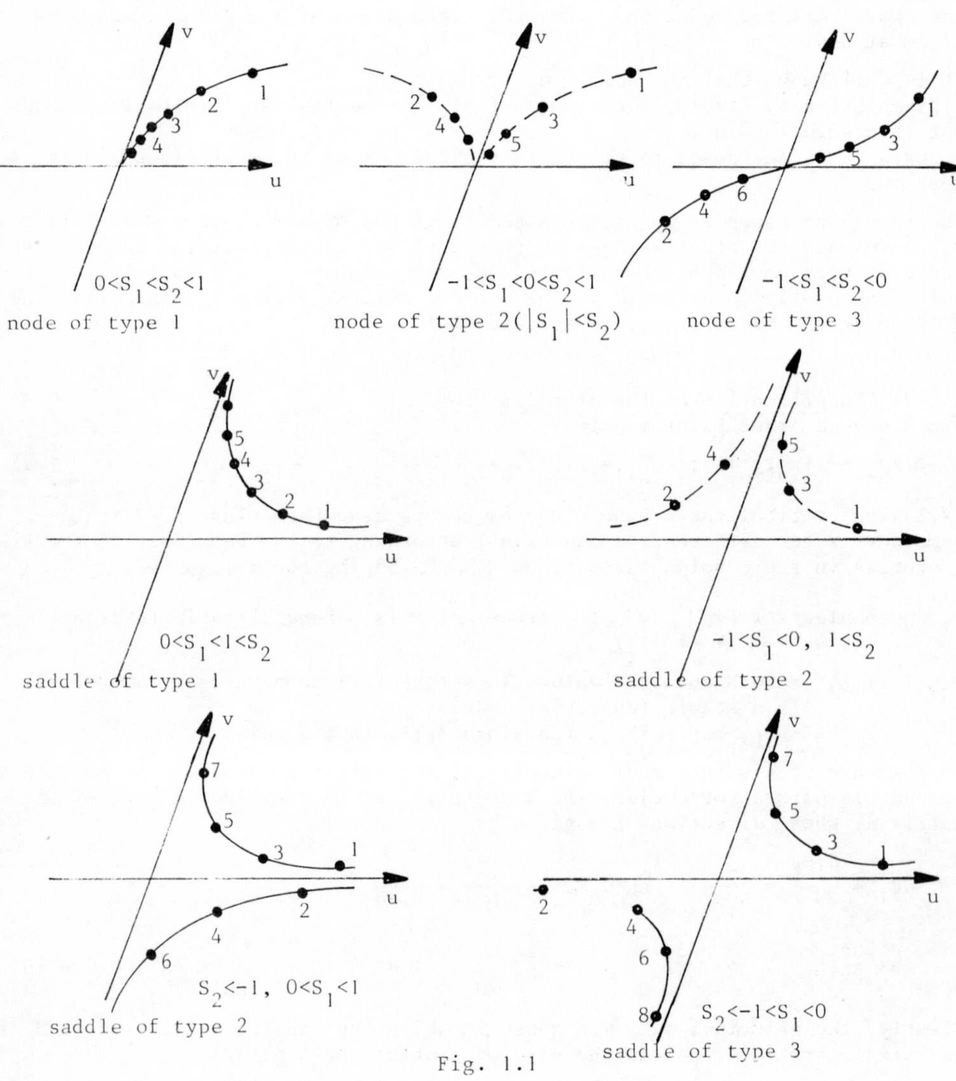

Fig. 1.1

Remarks:

a) Equal multipliers $S=S_1=S_2$ are only to be found in the node case, two canonical forms are possible :

$A=\begin{bmatrix} S & 0 \\ 0 & S \end{bmatrix}$; the principal axes are undetermined, the point is called a star node

$A=\begin{bmatrix} S & 1 \\ 0 & S \end{bmatrix}$; the principal axes are superimposed

b) The classification of 2nd order point mappings can be extended to higher by classifying the behaviour of the singularities in each invariant manifold corresponding to the different multipliers : several authors [eg ref. 3,p.126] have given such classifications for third order singularities which can still be represented geometrically.

c) Figure 1.2 shows the classical configuration in the Δ,σ plane (Δ=det[A], σ= trace[A]) for classification of double points or cycles.

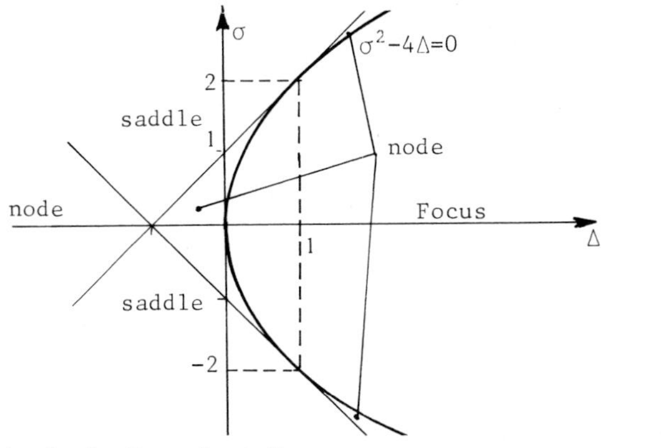

Fig. 1.2

1.2.3. Analytic Invariant Curves

If a point M_n is joined to the corresponding point M_{n+1} by means of a curve, then the iteration of this curve by the transformation or its inverse defines an invariant curve in the sense that every point of the curve is transformed into another point of the same curve. Among the innumerable invariant curves those which are of particular interest are the ones which pass through the singularities (node and saddle).

Let us consider the mapping(1.10), $S_1 \neq S_2$, the origin a double point and U and V non linear analytic forms, i.e., with convergent series expansions

$$U(u,v) = \sum_{k=2}^{\infty} \sum_{i=0}^{k} \left[\frac{\partial^k U}{\partial^2 u \partial^{k-i} v} \right]_0 \frac{u^i v^{k-i}}{i!(k-i)!} \tag{1.11}$$

$$V(u,v) = \sum_{k=2}^{\infty} \sum_{i=0}^{k} \left[\frac{\partial^k V}{\partial^i u \partial^{k-i} v} \right]_0 \frac{u^i v^{k-i}}{i!(k-i)!}$$

Hadamard (4) for the saddle case and Lattes (5) in general, showed that, provided that $S_1 - S_2^k \neq 0$ or $S_2 - S_1^k \neq 0$ for any positive integer k, there are two invariant curves C_1 and C_2, each tangential to a principal axis and analytical in a neigbourhood of the double point origin. Lattes also gives a straightforward way of determining the series expansions of the curves (C_1) and (C_2).

Considering the curve tangential to the u axis and its series expansions

$$v = H(u) \quad \sum_{i=1}^{\infty} h_i \frac{u^i}{i!} \tag{1.12}$$

the invariance of the curve can be written

$$S_2 H(u) + V(u,H(u)) = H[S_1 u + U(u,H(u))] \tag{1.13}$$

By identifying the coefficients of terms of the same power as u and v using (1.11) and (1.12) one obtains a set of equations which enable the h_i coefficients to be determined (theoretically). This can also be done by successive differentiation of equation (1.13) with respect to u, letting u=0 after each step. As a result one obtains

$$h_1 = 0 \, , \qquad h_2 = \left[\frac{\partial^2 V}{\partial u^2}\right]_0 (S_1^{\,2} - S_2)^{-1} \ldots$$

This method is rather cumbersome nor is it possible to calculate high degrees in the series expansions of (1.12). In chapter V other numerical methods are given to enable these invariant lines to be determined fairly accurately. It should be noted that these lines play an important role in defining the domain of influence of the singularities.

1.2.4. Other Definitions

In this section some of the problems discussed in the literature on point mapping but not covered in this book are mentioned.

1.2.4.1. Singularities at infinity. Among the equilibrium sets of the mapping those located at infinity must be included in order to give a complete qualitative study of the behavior. This can be done in the same way as for differential equations, [ref.2, p.324] using a method introduced by Poincaré (Poincaré's sphere) consisting of a central projection of the (x.y) plane onto a sphere of unit radius which is tangential to the (x.y) plane at the origin [ref.6,7].

1.2.4.2. Critical curves. For a second order point mapping, the critical curves are the loci of points which admit several antecedent superimposed. Then, such curves delimit zones in the (x.y) plane in which the number of antecedents of a point is constant.

1.2.4.3. Poincaré index. This notion has been extended by Neimark (8) to cover point mappings, in a very similar way to differential equations by considering the rotation properties of the vector $\overrightarrow{M_0M_1}$ (M_1 consequent of M_0) when M_0 describes a closed curve in the trigonometrical sense. The index is equal to : +1 if $\overrightarrow{M_0M_1}$ rotates by 2π, 0 — 0, -1 — -2π. Figure 1.3 gives the value of the index i in the $S_1.S_2$ plane when the closed curve described by M_0 encircles the singularity (node or saddle) only.

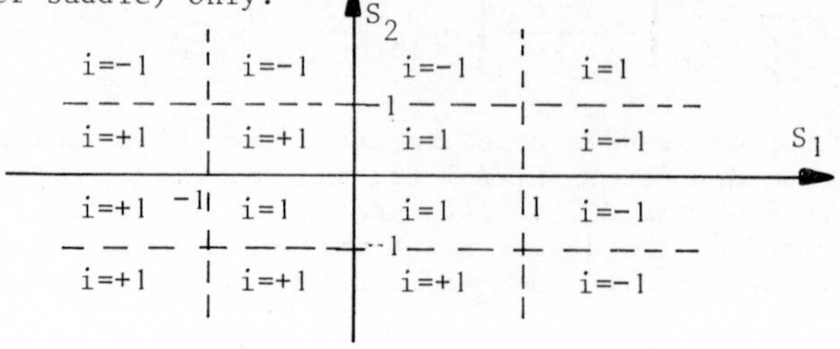

		S_2		
i=-1	i=-1	i=-1	i=1	
		1		
i=+1	i=+1	i=1	i=-1	S_1
i=+1	-1 i=1	i=1	1 i=-1	
		-1		
i=+1	i=+1	i=+1	i=-1	

Fig. 1.3

For a focus, i=+1. This can give some idea of the nature of the singularity without it being necessary to compute the eigenvalues

I.3. Some of the Main Problems Involved in Studying Point Mappings

The study of the solutions of non linear point mappings can cover an extremely broad area. In this book we have limited ourselves to autonomous non linear transformations. Insection 1.1 two typesof cases have been discussed :
1) Systems which can be directly described by means of recurrent equations
2) Systems to which a point mapping can be associated in order to provide information about their behaviour (chapter 3 gives an example of an application of this kind).
Let us now assume that we are faced with an autonomous point mapping

$$X_{n+1} = F(X_n) \tag{1.14}$$

An expression of this kind is clearly suited to numerical simulation, which in turn makes it worthwhile to look for the associated point mapping. Chapter 4 gives a method which in the case of conservative systems, enables an approximating associated mapping to be constructed, thus avoiding numerical integration. Some examples are given there. Although the first step of determining the mapping is time consuming once this is obtained the method is extremely economical.

A more systematic study of (1.14) would involve attempting to determine the singularities : double points and cycles. This involves the solution of a system of algebraic non linear equations, which is in fact an extremely difficult numerical problem. This is discussed briefly in chapter V. There a principle for determining the singularities is given based on the features of coarse systems. These are quite comprehensively covered by Andronov et al. (2).

Once the singularities have been obtained, the next step is to determine their nature by establishing their multipliers. In the case of a large scale system this is a somewhat delicate numerical problem and one which falls outside the scope of this work. Knowing the multipliers provides information regarding local stability, except in critical cases. Chapter II deals with some critical cases, whilst chapter IV is concerned with exceptional cases which are particular critical cases whose study is related to the so-called non linear parametric resonance phenomenon which arises for differential equations.

As noted above, knowing the multipliers of a double point or a cycle provides information regarding the local stability of these singularities as well as giving some idea of the qualitative phase topology of the solution in their neighbourhood. In order to understand the behaviour more fully, it is necessary to define the domain of influence of each singularity. A partial solution to this problem consists of extending Lyapunov's second method to equations of the type (1.14) (ref.8). Clearly, however this method gives only a part of the domain. In the case of 2nd order point mapping it is possible to adopt a more direct approach to find this domain.

To a stable limit set of (1.14) (double point, cycles of order m,...) corresponds a domain (D), composed of all the points whose consequents converge on this set. It should be noted that this domain should be non connected (ref.6). Let (F) be the boundary of the domain. One of the first contributions to this problem is found in the works of Julia and Fatou (9,10). They dealt with equation of the type

$$z_{n+1} = h(z_n) \ , \quad z \text{ complex} \tag{1.15}$$

which are a particular case of equations (1.14) where the right-hand members satisfy the Cauchy-Riemann conditions. They showed that (F) contains instable double points and cycles of all orders, of the star node type. In general, considering the boundary (F) of a stable equilibrium set, assuming (F) to be continuous, the following properties can be stated :

- the points of (F) form a closed set in the state plane and this constitutes a non attractive curve for the points in its neighbourhood
- (F) divides the plane into two domains each invariant with respect to the mapping and its inverse
- (F) is itself invariant with respect to the mapping and its inverse.

From these properties, if (F) contains double points or cycles :

- all the singularities are instable
- the antecedents of these singularities belongs to (F).

If one of these singularities is a saddle, the invariant curve corresponding to the multiplier less than one in absolute value belongs to (F). This enables in some cases, (F) to be determined precisely. Chapter V will give a more detailed discussion of this problem and an example.

Finally a further important problem is that of the bifurcation phenomenon, i.e. the sensitivity of the solution of (1.14) to parameter variations. Bifurcation can be defined as a change in the qualitative behaviour of the solutions of a recurrent equation, caused by a parameter variation. Among the numerous different possible cases of bifurcation those related to critical cases have been given the most attention (ref.2) and have led to the definition of dangerous and non dangerous stability boundary in the parameter space (ref.11). This problem is investigated in chapter V .

CHAPTER II
CRITICAL CASES

II.1 Introduction

A critical case in the sense of Lyapunov occurs for a double point or a cycle of a point mapping when, for the linear part of the equations, this singularity is stable but not asymptotically stable. The critical case is characterized by :

-at least, one multiplier, the absolute value of which is equal to 1
- the others which have absolute values strictly less than one.

In such a case,the stability analysis must take into account some of the terms of degree higher than one in the series expansions of the recurrent equations. It is clearly not possible at this stage to establish an uniform method of dealing with all the different types of critical cases. The complexity of the analysis is an increasing function of the number of critical variables (variables corresponding to multipliers whose absolute value is equal to one) appearing in the recurrent equations. A method involving non linear change of variable might be used, with the aim of determining a point mapping which is simpler in structure than the global mapping but which retains the same local properties for the stability problem.

The following section is devoted to the simplest critical case, i.e., one multiplier equal to one, the others with absolute value less than one. This will be useful for an introduction to some of the general ideas and methods that can be considered as basic to the study of critical cases. These have been mainly discussed in the work of Malkin (12). Section 2.4 deals with the critical cases with two multipliers of absolute value equal to 1. This involves considering :

- complex multipliers
- real multipliers, diagonal linear part
- real multipliers, non diagonal linear part.

Finally, some cases with more than two critical multipliers are considered.

II.2. Critical Case With One Multiplier Equal to One

The solution provided by Lyapunov (13) for differential equations can readily be used for recurrent equations. The following development is a brief recall of the results for the critical case with the multiplier equal to +1. The same solution can be used to deal with the negative multiplier case taking the second power of the mapping.

II.2.1. First Order Point Mapping

Let :

$$x_{n+1} = x_n + \sum_{i=m}^{\infty} a_i x^i , \qquad m \geq 2 \tag{2.1}$$

Theorem : the double point x=0 of (2.1) is asymptotically stable when m is odd and $a_m < 0$; and unstable in the other cases (m even or m odd and $a_m > 0$).

For m odd the Lyapunov function $V = x^2$ is chosen ;

$\Delta V = V(x_{n+1}) - V(x_n) = 2a_m x_n^{m+1} + \ldots$ has the sign of a_m around the double point.
For m even with the indefinite sign function $V(x)=x$ one gets $\Delta V = a_m x_n^m + \ldots$ which is sign definite in the neighbourhood of the origin. This makes it possible to establish instability of the double point origin using Lyapunov's first instability theorem, extended to the case of recurrent equations (Ref.8)

II.2.2. General Case, Some Remarks

Let us consider the $(\ell+1)$ dimensional mapping

$$x_{n+1} = x_n + f(x_n, Y_n)$$
$$Y_{n+1} = SY_n + G(x_n, Y_n) \quad (2.2)$$

where x is a scalar, Y an ℓ dimensional vector, f and G respectively an analytical function and an ℓ dimensional vector of analytical functions, series expansions of which begin with terms of at least second order with respect to their arguments. S is an $(\ell \times \ell)$ matrix with all eigenvalues having absolute value less than 1.
Let m be the minimum order of $f(x,0) = ax^m + \ldots$ higher order terms and p the minimum order appearing in the components of $G(x,0)$.

II.2.2.1. $m \leq p$. A Lyapunov function can be constructed as a sum of three terms $V = V_1 + V_2 + V_3$.

- V_1 being the Lyapunov function chosen for the first order point mapping $x_{n+1} = x_n + f(x_n, 0)$
- V_2 being a quadratic form, $Y^T BY$ with a definite matrix chosen in order to make ΔV sign definite
- V_3 being a form in terms of the (x,Y) variables which does not affect the sign definiteness of $V_1 + V_2$, coefficients of which are chosen to make ΔV sign definite.

For example if m is odd

$V_1 = 0.5\, x^2$

$V_2 = Y^T BY$, where B is a sign definite matrix chosen such that $S^T BS - B = aC$, C is a positive definite matrix. Then if $a < 0$, B is positive definite and $a > 0$, B is negative definite ($.^T$ means transpose).

$V_3 = \sum_{k=2}^{m} x^k C_k^T Y$ where C_k is an ℓ dimensional vector. V_3 has no influence on the sign definiteness of V in the neighbourhood of the origin since each of its terms contains x at least to the power 2. Clearly the C_k coefficients are chosen in such a way that are cancelled in the expressions of ΔV the terms, the degree of which is at least (m+1) with respect to x and 2 with respect to the components of Y.

This enables one to write

$$\Delta V = a\,(x_n^{m+1} + Y_n^T C Y_n) + R(x_n, Y_n)$$

R(.,.) having the above properties, V has the same sign as a in the neighbourhood of the origin. Thus :

a<0 ; V>0, ΔV<0 : the double point is asymptotically stable
a>0 ; ΔV<0 while V has a variable sign, the origin is unstable (Ref. 8)

If m even, the same method is available with : $V_1 = x$ (so that V will have avariable sign), V_2 is chosen in the same way as before and $V_3 = \sum_{k=1}^{m} x^k C_k^T Y$.

There is a slight modification which consists in starting the summation with k=1. This is due to the new form of the function V_1. Here also, $\Delta V = a(x_n^m + Y_n^T C Y_n) + R(x_n, Y_n)$ with fixed sign in the neighbourhood of the origin which is an instable fixed point since V has a variable sign (Ref. 8). It should be noticed that in such a case the stability property of the global mapping (2.2) is the same one as for the reduced first order point mapping

$$x_{n+1} = x_n + f(x_n, 0) \tag{2.3}$$

II.2.2.2. m>p. In this case the above method cannot be used easily, since the expressions of ΔV may contain terms depending on x_n only with degree less than (m+1) for m odd and (m) for m even. These terms may come from the function V_2. However this case can be reduced to the previous one by carrying out a non linear change of variables on (2.2) of the form :

$$Y = Z + U_L(x) = Z + \sum_{i=2}^{L} U_i(x), \text{ L integer} \tag{2.4}$$

where Z is an ℓ dimensional vector, and U_i a vector function consisting in homogeneous i^{th} degree polynomials.
In terms of the (x,Z) variables (2.2) is written

$$\begin{aligned} x_{n+1} &= x_n + f^*(x_n, Z_n) \\ Z_{n+1} &= SZ_n + G^*(x_n, Z_n) \end{aligned} \tag{2.5}$$

with

$$f^*(x,Z) = f(x, Z+U(x))$$

$$G^*(x,Z) = SU(x) + G(x, Z+U(x)) - U(x+f^*(x,Z))$$

Since eigenvalues of S have absolute value less than one it is possible (15) by means of a non linear change of variables to obtain $p^* = \min \text{degree}(G^*(x,0)) > M$ M integer. Elimination of the terms of $G^*(x,Z)$ depending on x alone is achieved by solving sequentially a set of equations of the form :

$$SU_i(x) - U_i(x) + G_i(x) + H_i(x) = 0 \quad , \qquad i=2,3,\ldots,M \tag{2.6}$$

where $G_i(x)$ are the ith degree terms in G(x,0) and $H_i(x)$ contains the ith degree terms of the series expansions of G(x,U(x)) and U(x+f(x+U(x))) other than $G_i(x)$ and $U_i(x)$ and which depend on $U_j(x)$ with j<i already determined.

This case can thus be reduced to the previous one if a coefficient a≠0 can be found. If not, the following case arises : $x_{n+1} = x_n + f(x_n, Z_n)$; $Z_{n+1} = SZ_n + G(x_n, Z_n)$ with f(x,0)=0 and C(x,0)=0 and the double point is stable (not asymptotically) (Ref. 13,16)
Remark : the condition p ≥ m is only a sufficient condition for applying the method of section II.2.2.1 and can be replaced by the following one :

"If during the iteration of (2.2), equations relative to Y do not introduce into the equation relative to x (the first of (2.2)) terms depending on x alone of degree less or equal to m then the problem can be solved by determining the stability of the first order equation : $x_{n+1} = x_n + f(x_n,0)$"

It is relatively simple, by looking at the expressions of the mapping, to see whether this condition is satisfied (at least for a reasonable order ℓ). This is the case for :

$$x_{n+1} = x_n + x_n^3 + y_n^2 x_n$$

$$y_{n+1} = Sy_n + x_n^2 \quad , \qquad S < 1$$

where m=3 and p=2 and there is no need for a change of variables. For a more detailed discussion see (Ref. 8,14)

II.3. Generalities. Malkin's Contribution

A survey of the works of Malkin is found in the books of Hahn (16,17) where he states :

"The general theory of the critical cases is quite involved and at this time the investigations have by no means been completed".

In our opinion, this is still the case, in particular for difference equations for which it has been given much less attention than for differential equations.

II.3.1. Notion Of Stability According To The m Degree Approximation

In the first chapter is given a list of the singularities which are stable according to the linear approximation (first degree approximation) i.e., singularities for which the local stability properties can be stated from the linear part of the expressions of the point mapping only. One of the fundamental contributions of Malkin was to generalise the concept of stability according to the first degree approximation by introducing the notion of stability according to the m degree approximation.

A singularity (fixed point or cycle) of a point mapping is said to be stable (or unstable) according to the m degree approximation (or, more simply, in the m degree approximation) whenever the stability (or instability) property is independent of the terms of degree higher than m.

In what follows, unless indicated otherwise, the capital letters X,Y will denote vectors, and small ones scalars. Consider the mapping :

$$X_{n+1} = F_m(X_n) + F_{m+1}(X_n) \ ; \quad F_m(0) = F_{m+1}(0) \equiv 0 \qquad (2.7)$$

where $F_m(X_n)$ is composed by the terms of the series expansions of the mapping, the degree of them is less than or equal to m, $F_{m+1}(X_n)$ represents the remaining terms with the condition $\| F_{m+1}(X_n) \| < a \, \| X_n \|^{m+1}$, a constant. This implies that, at least, the mapping has a Taylor expansion of up to m degree. Generally, for convenience, it is assumed that the mapping is analytic.

The fixed point origin of the mapping is stable in the m degree approximation if :

$$\forall \varepsilon > 0, \ \exists \, \delta(\varepsilon,a) \quad \text{such that if } \| X_n \| < \delta \Rightarrow \| X_{n+\ell} \| < \varepsilon, \qquad \forall \ell \text{ integer} > 0$$

Under the above conditions, it is obvious that a mapping which is stable (unstable) in the mth approximation is stable (unstable) in the ℓth approximation for each $\ell \geq m$. For practical purposes, it is useful to know (or to find) the minimum degree m that will enable the stability problem to be solved. In what follows, it is this minimum degree that is referred to implicitly, when speaking of stability in the m degree approximation. This implies, in addition, that the mapping is "critical" up to the (m-1) order terms.

II.3.2. Dimensionality Reduction

Let us consider the general problem of an ℓ dimensional point mapping, the origin being a fixed point. First, by means of a linear change of variables, it is possible to separate the linear part of the mapping into two parts ; the first depending only on the critical variables (that is the variables associated with the critical multipliers), the second on the asymptoticaly stable variables. The mapping is then expressed :

$$\begin{aligned} X_{n+1} &= CX_n + F(X_n, Y_n) \\ Y_{n+1} &= DY_n + G(X_n, Y_n) \end{aligned} \tag{2.8}$$

where dim(X)=p, dim(Y)=q ($p+q=\ell$) and C,D are matrices for which the eigen values have absolute values equal to one and less than one, respectively.

In the preceding section, where p=1, the stability study of the global system (2.8) was reduced to that of a first order system, i.e., of order equal to the number of critical variables. This approach can be used in other cases. Hahn (17, p.112) suggests some way of handling the critical case of equation (2.8). The first step consists on applying certain non linear transformations to the non critical variables (Y) in order to reduce the dependence of the corresponding equations with respect to the critical variables by eliminating terms of degree less than m, for any m. The aim is to construct a mapping in which the non critical variables have no effect upon the stability of the equilibrium, which can be determined by studying a reduced mapping of order equal to the number of critical variables. This is generally done by constructing a suitable Lyapunov function for this reduced mapping.

Other works on differential equations which should be mentioned here are those of (i) Zubov (18) who dealt with the cases involving several zero or imaginary roots and (ii) more recently, Oziraner (19) who has studied the critical case with both zero and imaginary eigenvalues. For difference equations it is worth mentionning the works by Simanov-Kazeeva (Ref. 35)

II.4. Critical Cases With Two Multipliers Of Absolute Value Unity

The section is divided into two parts, the first dealing with complex multipliers case ; the second real multipliers case.

II.4.1. Complex Multipliers

II.4.1.1. Second order point mapping (Ref. 20,21). Consider the real mapping

$$\begin{aligned} r_{n+1} &= r_n \cos w - s_n \sin w + f(r_n, s_n) \\ s_{n+1} &= r_n \sin w + s_n \cos w + g(r_n, s_n) \end{aligned} \tag{2.9}$$

with $f(0,0)=g(0,0)=0$, for which multipliers are $\exp[\pm jw]$.

For the sake of convenience in the following calculations the change of variables $x = r+js$, $y = r-js$ is performed on (2.9) and the transformed mapping (with complex variables and coefficients) is written

$$x_{n+1} = \exp[jw]\, x_n + \sum_{k+\ell=2}^{\infty} C_{k\ell} x_n^k y_n^\ell$$

$$y_{n+1} = \exp[-jw] y_n + \sum_{k+\ell=2}^{\infty} C_{k\ell}^{*} x_n^\ell y_n^k \tag{2.10}$$

where * means conjugate. The linear part of (2.9) representing a rotation of angle w makes it necessary, to try finding a Lyapunov function of the form :

$$V = r^2 + s^2 + \sum_{m+n=3}^{\infty} v_{mn} r^m s^n$$

that is in complex variables

$$V(x,y) = xy + \sum_{m+n=3}^{\infty} a_{mn} x^m y^n \ , \quad a_{mn} = a_{nm}^{*} \tag{2.11}$$

The idea is to express the increment $\Delta V = V(x_{n+1}, y_{n+1}) - V(x_n, y_n)$ in a sign definite form by a proper choice of the a_{mn} coefficients. For instance, looking at the third order terms of ΔV one gets

$$E_3 = \sum_{k+\ell=2} \exp(-jw) C_{k\ell} x_n^k y_n^{\ell+1} + \exp(jw) C_{k\ell}^{*} x_n^{k+1} y_n^\ell) +$$

$$\sum_{k+\ell=3} (\exp(j(k-\ell)w) - 1)\ a_{k\ell} x_n^k y_n^\ell$$

Provided that $\exp[j(k-\ell)w] - 1 \neq 0$ whatever k, ℓ with $k+\ell=3$ every term of E_3 can be cancelled by a suitable choice of the a_{mn} parameters. The above condition excludes the following cases :

$w = 2k\pi(m-n)^{-1}$, with m,n positive integers and m+n=3, k integer, i.e.,

$w = 2k\pi$, corresponding to the critical case with two real multipliers (this will be treated in the next section)

$w = \frac{2k\pi}{3}$, this will be called the "exceptional case" for the mapping. It can be reduced to the study of the critical case with two multipliers equal to one. This will be described in chapter IV in the section about parametric resonance occuring in some non linear differential equations with periodical coefficients.

For the fourth degree terms, we have

$$E_4 = \sum_{k+\ell=3} (\exp[-jw] C_{k\ell} x_n^k y_n^{\ell+1} + \exp[-jw] C_{k\ell}^{*} x_n^{k+1} y_n^\ell) + \sum_{k+\ell=4} (\exp[j(k-\ell)w] - 1)$$

$$a_{k\ell} x_n^k y_n^\ell + R_4(x_n, y_n)$$

where $R_4(x_n, y_n)$ contains the fourth degree terms (other than those written explicitely) appearing in the expressions of ΔV and depending on $a_{k\ell}$ coefficients (with $k+\ell=3$) which have already been determined by the first step of the elimination. If the exceptional case is excluded ($w \neq 2k\pi/q$, q integer) it is clear that the terms of E_4 can be cancelled, except that for which :

$k = \ell = 2$

Then,

$$\Delta V = G_4 (x_n y_n)^2 + \ldots = G_4 (r_n^2 + s_n^2)^2 + \ldots \tag{2.12}$$

where the dotted lines represent terms of degree higher than 4. It follows that :

$G_4 < 0$ (>0)=> double point asymptotically stable (unstable)

In this case, using the terminology of section 2.3 it can be said that the double point is stable (unstable) in the third approximation. If $G_4=0$, it is not possible to study stability using the terms of third degree or less. The calculations must be carried out taking into account the 5th and 6th degree terms of the series expansions V (2.11). The elimination of 5th and 6th degree terms of ΔV involves expressions like E_4 in which : $\exp[j(k-\ell)w]-1 = 0$. Excluding $w= 2k\pi/q$ the above relation is satisfied only by the terms for which $k=\ell$. Thus, all odd degree terms can be cancelled whereas among the even degree terms, the terms like $(x_n y_n)^k$ remain. For instance, in terms of degree six, ΔV is written

$$\Delta V = G_6(x_n y_n)^3 + \ldots = G_6(r_n^2 + s_n^2)^3 + \ldots$$

and the stability is fixed by the sign of G_6.

In conclusion, if $G_{2\ell}$, ℓ integer >0 is the first non null coefficient appearing in the expression of ΔV, the double point (also called focus of multiplicity ℓ) is stable (unstable) if $G_{2\ell}<0$ (>0). In this case there is stability (instability) in the $(2\ell-1)^{th}$ approximation.
For area preserving mapping (conservative system) $G_{2\ell}= 0, \forall \ell$.

II.4.1.2. General case (Ref. 22,23). In this section we outline some of the basic considerations involved in defining a Lyapunov function for the general mapping :

$$\begin{aligned} r_{n+1} &= r_n \cos w - s_n \sin w + f(r_n, s_n, Y_n) \\ s_{n+1} &= r_n \sin w + s_n \cos w + g(r_n, s_n, Y_n) \\ Y_{n+1} &= AY_n + H(r_n, s_n, Y_n) \end{aligned} \tag{2.13}$$

It is assumed that the mapping :

$$\begin{aligned} r_{n+1} &= r_n \cos w - s_n \sin w + f(r_n, s_n, 0) \\ s_{n+1} &= r_n \sin w + s_n \cos w + g(r_n, s_n, 0) \end{aligned} \tag{2.14}$$

is stable or unstable in the $(2\ell-1)$ approximation and that the minimum degree of $H(r_n, s_n, 0) \geq 2\ell-1$. The Lyapunov function is chosen as the sum of three terms

$$V = V_1 + V_2 + V_3$$

where

- V_1 is the Lyapunov function chosen for the second order mapping (2.14) i.e., $V_1 = r^2 + s^2 + \ldots$ and $\Delta V_1 = G_{2\ell}(r_n^2 + s_n^2)^\ell$
- V_2 is a definite quadratic form $Y^T BY$ such that $A^T BA - B = G_{2\ell} D$ D being a positive definite matrix. B is positive (negative) definite for $G_{2\ell}<0$ (>0)
- V_3 is made up of a sum of terms such as $r_n^i s_n^j D_{ij}^T Y_n$ with $i+j=3,4,\ldots 2\ell$, D_{ij} is a vector, the components of which are chosen to enable the terms of ΔV of degree less than ℓ with respect to r_n^2 and s_n^2 and 2 with respect to Y_n to be cancelled. It is easy to see that V_3 has no influence upon the sign definiteness of the function V.

Then, $\Delta V = G_{2\ell}[(r^2 + s^2)^\ell + Y^T DY] + \ldots$

ΔV is of sign definite in the neighbourhood of the origin. If $G_{2\ell}<0$, V is positive definite, 0 is asymptotically stable ; $G_{2\ell}>0$, V is undefined and 0 is unstable.

In appendix 1 the expression of G_4 is given, which gives some idea of the complexity of the calculations.

Remark : an other way of tackling the critical case is to perform on (2.9) or (2.10) a non linear change of variables, called Cigala substitution (24) of the form (in complex variables)

$$\bar{x} = x + \sum_{i+j=2}^{L=2\ell-1} a_{ij} x^i y^j \qquad \ell \text{ integer } >0 \qquad \bar{y} = y + \sum_{i+j=2}^{L} a_{ij}^{**} x^j y^i \tag{2.15}$$

in order to get a simpler mapping in terms of the $(\bar{x},\bar{y})$ variables, by cancelling, in these expressions, the maximum number of terms by a proper choice of the a_{ij} coefficients (Ref. 20). When non exceptional case arises, the next mapping is written

$$\bar{x}_{n+1} = \exp[jw]\bar{x}_n + \sum_{k=1}^{\ell} \bar{C}_{k+1,k} \bar{x}_n^{k+1} \bar{y}_n^{k} + \ldots \qquad \bar{y}_{n+1} = \exp[-jw]\bar{y}_n + \sum_{k=1}^{\ell} \bar{C}_{k+1,k}^{**} \bar{x}_n^{k} \bar{y}_n^{k+1} + \ldots \tag{2.16}$$

for which the Lyapunov function $V=\overline{xy}$ can be used.

II.4.2. Real Multipliers. Linear Diagonal Part

If the two critical multipliers are equal to one, the negative multipliers case can be dealt with in the same way as the positive one by taking the 2nd power of the mapping. In this section a 2nd order mapping with a diagonal linear part is considered.

$$x_{n+1} = x_n + f(x_n,y_n) \qquad y_{n+1} = y_n + g(x_n,y_n) \qquad f(0,0)=g(0,0)=0 \tag{2.17}$$

II.4.2.1. Assumptions. Preliminary transformations. For convenience, f and g are assumed to be analytic in a neighbourhood of the origin and their series expansions begin with the same degree m

$$f(x,y)=\sum_{j=m}^{\infty} f_j(x,y) \quad ; \quad g(x,y)=\sum_{j=m}^{\infty} g_j(x,y)$$

with $m\geqq 2$, f_j, g_j homogeneous polymonials of degree j.

It is not worth trying to find a Lyapunov function of the same type as that described in the preceding section in particular since the a_{mn} coefficients disappear naturally in ΔV. This arises from the fact that the linear part of (2.17) is an identity (not a rotation as before). If the polar coordinate change x=rcosw, y=rsinw is carried out, then (2.17) can be written

$$r_{n+1}^2 = r_n^2 + P_m(w_n) r_n^{m+1} + \ldots \qquad w_{n+1} = w_n + Q_m(w_n) r_n^{m-1} + \ldots \tag{2.18}$$

where the dotted lines represent terms of higher degree than (m+1) and (m-1) respectively with respect to r and where :

$$
\begin{aligned}
P_m(w) &= 2(\cos w\, f_m(\cos w, \sin w) + \sin w\, g_m(\cos w, \sin w)) \\
Q_m(w) &= \cos w\, g_m(\cos w, \sin w) - \sin w\, f_m(\cos w, \sin w)
\end{aligned}
\tag{2.19}
$$

Two cases are considered :

(i) $Q_m(w)$ has a definite sign for $w\in[0,2\pi]$ (there is no invariant curve passing through the origin)

(ii) $Q_m(w)$ does not have a definite sign for $w\in[0,2\pi]$ (there may be invariant curves through the origin).

II.4.2.2. $Q_m(w) \neq 0$, $w\in[0,2\pi]$. The double point origin is a focus type (or center) point. The stability problem can be solved using the Lyapunov function.

$$
V = r \exp\left[\int_0^w h(u)\, du\right] \tag{2.20}
$$

where $h(u)$ is a periodic function of u of period 2π with null mean value. The increment ΔV is written

$$
\begin{aligned}
\Delta V &= r_{n+1} \exp\left[\int_0^{w_{n+1}} h(u)\, du\right] - r_n \exp\left[\int_0^{w_n} h(u)\, du\right] \\
&= r_n^m \exp\left[\int_0^{w_n} h(u)\, du\right]\left[\frac{P_m(w_n)}{2} + h(w_n)Q_m(w_n) + r\ldots\right]
\end{aligned}
$$

Defining $h(w) = h_0 - \dfrac{P_m(w)}{2Q_m(w)}$ with $h_0 = \dfrac{1}{2\pi}\displaystyle\int_0^{2\pi} \dfrac{P_m(w)}{2Q(w)}\, dw$

we get

$$
\Delta V = r_n^m \exp\left[\int_0^{w_n} h(u) du\right] \cdot h_0 Q_m(w_n) + r_n^{m+1} \ldots \tag{2.21}
$$

The double point origin is stable (unstable) if $hQ_m(w)<0(>0)$. If $h_0 = 0$ the m degree approximation is not sufficient to determine stability and higher degree terms are needed. Appendix 2 gives the solutions for this case (Ref. 25)

II.4.2.3. $Q_m(w)$ of sign not defined. Invariant analytical curves may pass through the double point. For the truncated system

$$
\begin{aligned}
x_{n+1} &= x_n + f_m(x_n, y_n) \\
y_{n+1} &= y_n + g_m(x_n, y_n)
\end{aligned}
\tag{2.22}
$$

it is easy to verify the fact that the straight lines D_i,

$$
D_i : y - p_i x = 0 \quad ; \quad p_i = \mathrm{tg}(w_i) \quad ; \quad Q_m(w_i) = 0
$$

are in fact invariant by the mapping (2.22). Then the following theorem holds: If $P_m(w_i)<0$, $\forall w_i : Q_m(w_i)=0$ the double point is asymptotically stable. If $\exists\, w_i$: $Q_m(w_i)=0$ and $P_m(w_i)>0$ the double point is unstable.

The second part of the theorem is quite obvious since $P_m(w_i)>0$ means that on the invariant line the sequence of iterates diverges from the double point (increment of $V= r^2+s^2$). Before demonstrating the first part of the theorem, using the same approach done by Nemitski-Stepanov (Ref. 26) for differential equation the invariant lines are classified according to the behaviour of the discrete trajectories in their neighbourhood. It is assumed that $Q_m(w_i)\neq 0$. Let w_i be a root of multiplicity n_i.

$$Q_m(w) = (\sin w - p_i \cos w)^{n_i} Q_{m-n_i}(w) \;, \; p_i = \mathrm{tg} w_i \tag{2.23}$$

A sector $[\bar{w}_{i-1}, \bar{w}_{i+1}]$ can be defined where $P_m(w)$ has the same sign (assumed to be negative) and which contains only the root w_i. Two cases can be considered, depending on n_i.

(i) n_i odd. If $Q_{m-n_i}(w_i)>0$ the line $w=w_i$ is repulsive and the discrete trajectories behave as shown in Fig. 2.1.

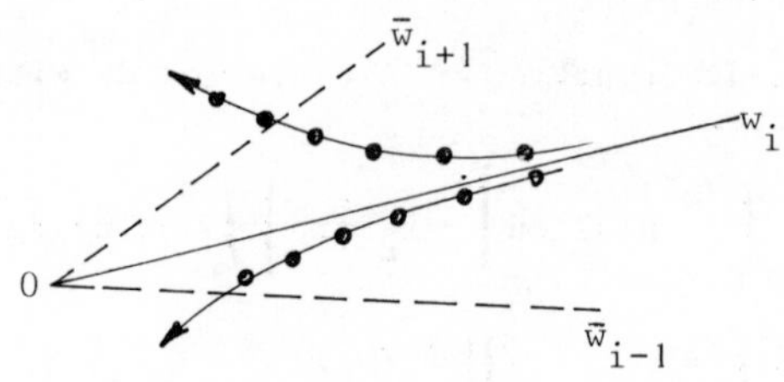

Fig.2.1

Inversely, if $Q_{m-n_i}(w_i)$ is negative, the line $w = w_i$ is attractive and the discrete trajectories are tangential, at the origin, to the line $w = w_i$ (Fig.2.2).

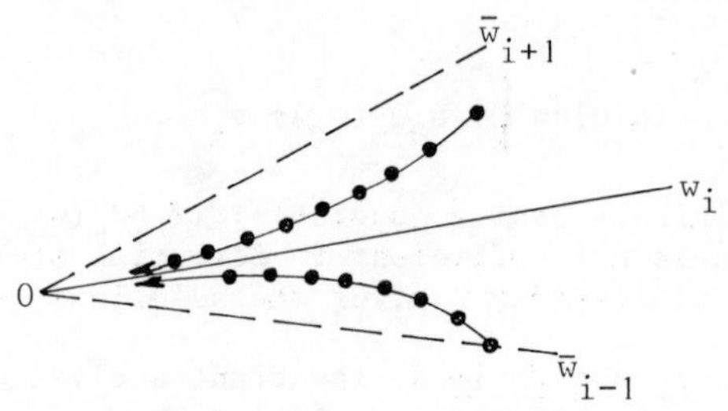

Fig.2.2

It should be noted that for $n_i=1$ these two conditions correspond respectively to $[\frac{dQ_m}{dw}]_{w_i} > 0$ and $[\frac{dQ_m}{dw}]_{w_i} < 0$ (Ref. 27).

(ii) n_i even. The behaviour of the solutions is not the same on both sides of $w = w_i$, Fig.2.3.

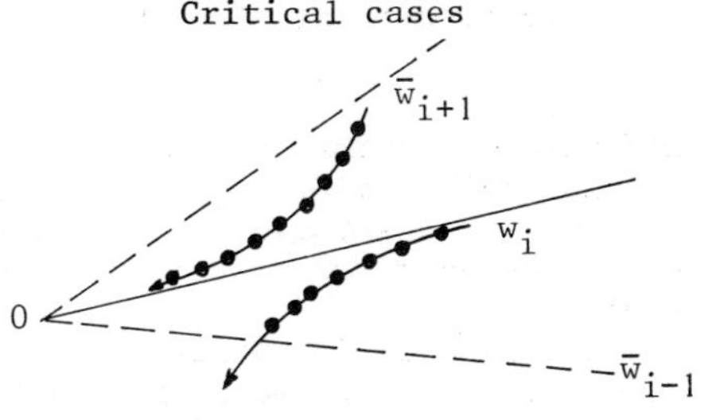

Fig. 2.3

The demonstration of the first part of the theorem is made by considering the behaviour of the solutions in a sector defined by two adjacent invariant directions w_i and w_{i+1} :

$$Q_m(w_i)=0 \;;\quad Q_m(w_{i+1})=0 \quad \text{and} \quad Q_m(w)\neq 0 \quad \forall w\in[w_i,w_{i+1}]$$

If $Q_m(w)<0, \forall w\in[w_i,w_{i+1}]$ r decreases monotically and stability obviously holds in the sector. If such a condition is not fulfilled, there exist nevertheless two sub sectors $[w_i,\bar{w}_i],[\bar{w}_{i+1},w_{i+1}]$ in which $P_m(w)<0$.

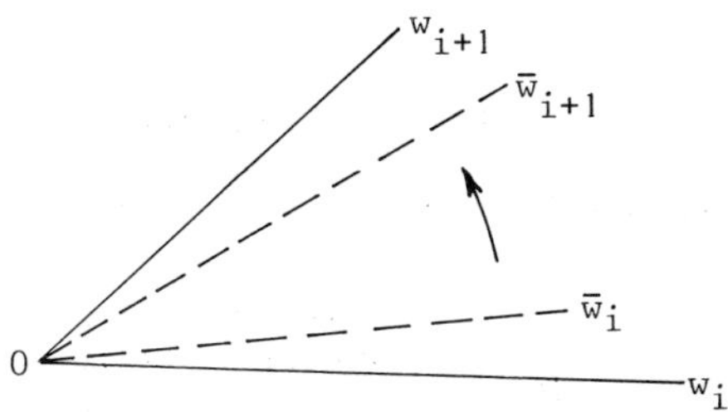

Fig. 2.4

For $Q_m(w)>0$, $w\in[w_i,w_{i+1}]$ the iterates move in the direction indicated by the arrow. The problem is to demonstrate that the iterates of the mapping (2.22) cannot diverge in the sector $[w_i,w_{i+1}]$.

The mapping (2.22) is stable (in the sector) if :

$$\forall\delta>0, \exists\,\varepsilon>0 : r_n^2<\varepsilon \;\Rightarrow\; r_{n+k}^2<\delta \quad , \forall k \text{ integer } >0$$

Let $A=\max_w P_m(w)$, $a=\min_w Q_m(w)$, $w\in[\bar{w}_i,\bar{w}_{i+1}]$

Whatever the initial condition in a sufficiently small neighbourhood of the origin and in the sector $[w_i,w_{i+1}]$. It can be shown that :

$$r_{n+k}^2 < r_n^2 \exp\,[A(w_{n+k}-w_n)a^{-1}] \tag{2.24}$$

so, $\forall\delta>0, \exists\,\varepsilon>0$ satisfying the stability condition : eg $\varepsilon=\delta\exp^{-1}[A(\bar{w}_{i+1}-\bar{w}_i)a^{-1}]$

In the preceding section the study was related to the truncated mapping (2.22) the m degree approximation of the mapping. As stated in section 2.3 it can be shown that in all cases when the answers to the stability problem are unambigous the stability (or instability) found for (2.22) apply for the total mapping as well. These conditions will be summarised at the end of the present section.

Remark : (i) In the case when $Q_m(w)$ does not have a definite sign the problem arises of the nature of the double point origin : i.e., the problem of the phase portrait of the discrete trajectories around this point. Under what conditions do invariant curves pass through the origin (tangential to the invariant straight lines of (2.22))?
This is a complex problem dealt with for differential equations in the book of Nemitski -Stepanov. Here it is not treated in full details. The existence of invariant curves through the origin essentially depends on the parity of multiplicity of the roots of $Q_m(w)=0$, that is on n_i. This is clear from the following figures:

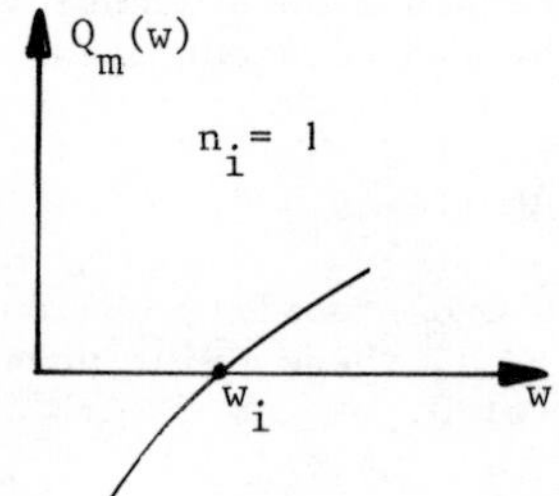

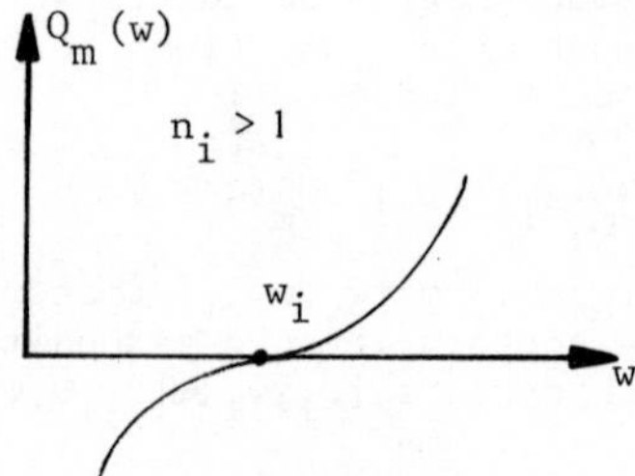

Fig. 2.5 - n_i odd

An invariant curve, tangential at the origin of the direction $w=w_i$ exists

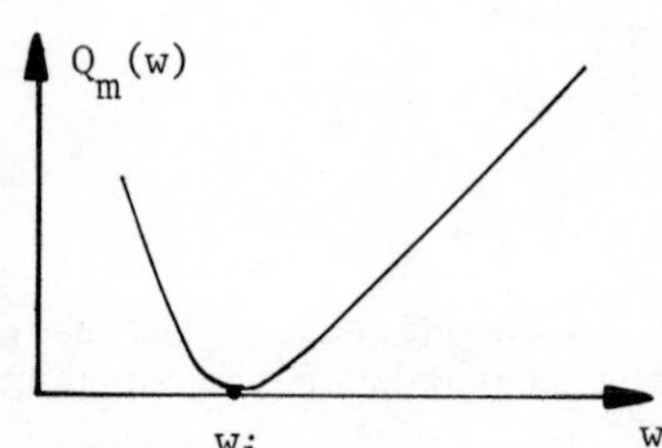

Fig. 2.6 - n_i even

There is uncertainty so that higher degree terms of the series expansions need to be taken into account. There may be no, one or several invariant curves. Below the possible phase portraitsfor $n_i = 2$ ($P(w_i)<0$) are shown.

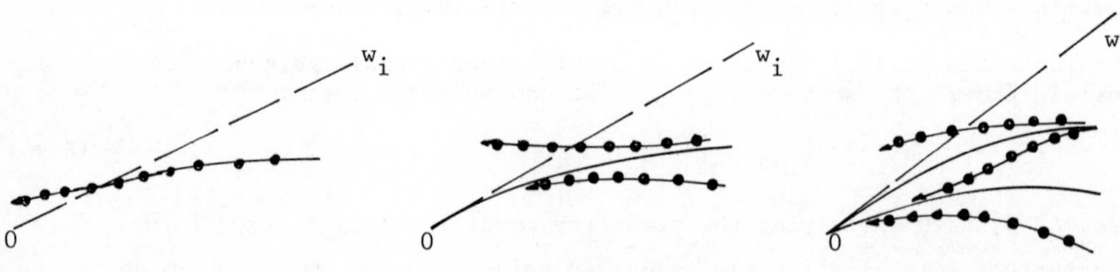

no invariant curve one invariant curve two invariant curves

Fig. 2.7

(ii) Equation $Q_m(w)=0$ is of degree (m+1) with respect to p=tgw. Thus there will be at the most (m+1) invariant directions, defining (m+1) sectors. It is clear that there are three types of sectors :

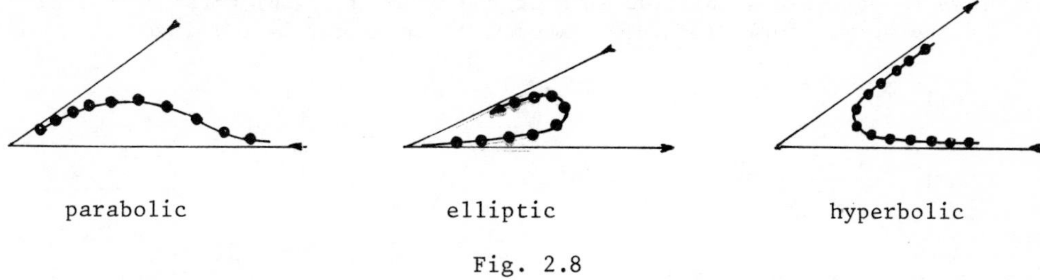

Fig. 2.8

II.4.2.4. Summary. For

$$x_{n+1} = x_n + \sum_{j=m}^{\infty} f_j(x_n,y_n)$$
$$y_{n+1} = y_n + \sum_{j=m}^{\infty} g_j(x_n,y_n) \qquad (2.25)$$

with f_j, g_j homogeneous polynomials of degree j and

$$P_m(w) = \cos w\, f_m(\cos w, \sin w) + \sin w\, g_m(\cos w, \sin w)$$

$$Q_m(w) = \cos w\, g_m(\cos w, \sin w) - \sin w\, f_m(\cos w, \sin w)$$

two cases are encountered

(i) $Q_m(w) \neq 0,\ w \in [0,2\pi]$ $\quad h_0 = \int_0^{2\pi} \frac{P_m(w)}{Q_m(w)} dw$. The origin (double point) is asymptotically stable (unstable) if $h_0 Q_m(w)<0$ $(h_0 Q_m(w)>0)$.

(ii) $Q_m(w)= 0$ for $w=w_0, w_1, \ldots, w_\ell, \ell \leq m$. The origin is asymptotically stable if $P_m(w_i)<0, \forall i=0,1,..\ell$; unstable if $\exists$ i so that $P_m(w_i)>0$

When $h_0 = 0$ and $P_m(w_i)= 0$ higher order terms than m need to be taken into account, some solutions are given in appendix 2.

II.4.3. Non Diagonal Linear Part

Here again both multipliers are equal to +1, the linear part of the mapping being non diagonal

$$r_{n+1} = r_n + s_n + f(r_n,s_n)$$
$$s_{n+1} = s_n + g(r_n,s_n) \qquad f(0,0)=g(0,0)=0 \qquad (2.26)$$

where f and g are analytic functions whose series expansions begin with terms of degree at least equal to 2.

II.4.3.1. Preliminary transformations. Normal forms. Unlike to the previous critical case it is possible to perform on (2.26) a non linear transformation of the form

$$r = x + \sum_{i+j=2}^{M} a_{ij} x^i y^j \ , \quad s = y + \sum_{i+j=2}^{M} b_{ij} x^i y^j \tag{2.27}$$

in order to obtain a simplified form of the mapping (2.26) this form being called the normal form (Ref. 15). Two normal forms can be obtained

$$\begin{aligned} x_{n+1} &= x_n + y_n + \sum_{j=m}^{M} C_{j0} x_n^j + E_1(x_n, y_n) \\ y_{n+1} &= y_n + \sum_{j=n}^{M} D_{j0} x_n^j + E_2(x_n, y_n) \end{aligned} \qquad m,n \geq 2 \tag{2.28}$$

and

$$\begin{aligned} x_{n+1} &= x_n + y_n + E_3(x_n, y_n) \\ y_{n+1} &= y_n + \sum_{j=n'}^{M} D_{j0} x_n^j + \sum_{j=m'-1}^{M-1} D_{j1} x_n^j \, y_n + E_4(x_n, y_n) \end{aligned} \qquad m',n' \geq 2 \tag{2.29}$$

where E_1, E_2, E_3, E_4 are analytic functions whose series expansions do not include terms of degree lower than M+1. Because of the very numerous subcases which may be encountered, we will restrict ourselves to a general survey of the methods used for studying the stability problem. The reader is referred to (Ref. 28) for differential equations and to (Ref. 15) for point mapping for fuller details and a summary of the results for the various subcases.

II.4.3.2. Invariant curves. One way of dealing with stability is to determine under what conditions invariant curves pass through the origin. A curve $y=h(x)$ is invariant under (2.28) if it satisfies the following functional equation

$$h[x + h(x) + f(x,h(x))] = f(x) + g(x,h(x)) \tag{2.30}$$

i.e., $h(x_{n+1}) = y_{n+1}$ whenever $h(x_n) = y_n$. Here $f(x_n, y_n)$ and $g(x_n, y_n)$ denote the non linear terms of x_{n+1} and y_{n+1} respectively as expressed in (2.28). An inspection of the expressions (2.28) shows that h(x) may be expressed as a series expansion of the form

$$h(x) = p_0 x^a + \sum_{i=1}^{\infty} P_i x^{a+i/2} \ , \quad \text{i positive integer} \tag{2.31}$$

where $a = k/2 > 1$, k integer. Identifying the lower degree terms of both members of (2.30) taking (2.31) into account, one gets the necessary conditions for the existence of real invariant curves. Three cases will be considered here :

case A - $(n<2m-1)$, $a=(n+1)/2$, $p_0^2 = 2D_{n0}/(n+1)$

case B - $(n=2m-1)$, $a=m$, $p_0 = (-C_{m0} \pm \sqrt{C_{m0}^2 + 4D_{n0}/m})/2$

case C - $(n>2m-1)$, $a=n-m+1$, $p_0 = D_{n0}/(C_{m0}(n-m+1))$

or $a = m$, $p_0 = -C_{m0}$

All the coefficients P_i in (2.31) can be determined as functions of the coefficients of (2.28) and P_j with $j<i$. However in case B an additional condition must be satisfied for this to be true

$$(2m+q)p_0 \neq -(m+q)C_{m0} \qquad \forall q \text{ integer } >0 \tag{2.32}$$

The conditions for the realness of such curves are :

case A - $D_{n0}>0$ if n is odd, any $D_{n0} \neq 0$ if n is even

case B - $C^2_{m0}+4D_{n0}/m>0$

case C - The curves are always real.

These conditions follow immediatly from the above expressions for p_0 and from the fact that if p_0 is real then $p_j (j=1,2,..)$ is also real. (Note that in case A if n is even D_{n0} may be assumed to be positive, the substitution $x \to -x$, $y \to -y$ may be performed if necessary).

In (Ref.29) the convergence of the series (2.31) is demonstrated in the neighbourhood of the origin. It should be noticed that in the case B with $C^2_{m0}+4D_{n0}/m>0$, one real analytic invariant curve always exists independently of the restriction (2.32) (Ref. 15).

Having found such invariant curves some partial answers to stability problem can be provided. Let us denote by T the point mapping defined by (2.28). Then an invariant curve (2.31) is said to be an Ω curve or an α curve if a point belonging to it in a sufficiently small neighbourhood of the origin tends to the origin under successive applications of T or T^{-1}, respectively. Since the existence of an α curve is sufficient to guarantee the instability of the origin we can derive some instability criteria. If all curves (2.31) are Ω curves overall stability must be demonstrated as in section II.4.2.3. The cases treated below include this particular one and also those which are characterised by the absence of real invariant curves of type (2.31).

II.4.3.3. Other cases. Here we shall give only a brief description of the method used to deal with these cases. This method (which is described more fully in the following chapter) consists essentially of constructing a differential equation which generates the mapping (2.26) or (2.28) (see theorem below). The Lyapunov's results (28) may then be readily extended to this class of differential equations.

In (30, p.113-125) Birkhoff showed that the series x_k, y_k defining the kth consequent of the point x,y under (2.28), i.e.,

$$\begin{aligned} x_k &= x + ky + \sum_{i+j=m}^{\infty} C_{ij}^{(k)} x^i y^j \\ y_k &= y + \sum_{i+j=n}^{\infty} D_{ij}^{(k)} x^i y^j \end{aligned} \qquad k \text{ integer, } m \leqq n \tag{2.33}$$

satisfies the formal differential system (k being now considered as a continuous variable)

$$\frac{dx_k}{dk} = U(x_k, y_k) \qquad \frac{dy_k}{dk} = V(x_k, y_k) \tag{2.34}$$

where $U(x,y) = \left[\frac{\partial x_k}{\partial k}\right]_{k=0}$ $\qquad V(x,y) = \left[\frac{\partial y_k}{\partial k}\right]_{k=0}$

and $x_0 = x \qquad y_0 = y$

In particular for the mapping (2.28), elementary calculations lead to a system (2.34) of the form :

$$\frac{dx}{dt} = y + \sum_{i+j=m}^{\infty} E_{ij}x^i y^j$$
$$\frac{dy}{dt} = \sum_{i+j=n}^{\infty} F_{ij}x^i y^j \tag{2.35}$$

where we assume $n \geqq m$. It is easy to see that if $n<m$ then the sum in the expression for $\frac{dx}{dt}$ begins with $i+j=n$. Therefore we may assume $m \leqq n$ without loss of generality for the results obtained.

Theorem : It is possible to define a differential system with periodic coefficients of the form

$$\frac{dx}{dt} = y + \sum_{i+j=m}^{L} E_{ij}x^i y^j + H_1(x,y,t)$$
$$\frac{dy}{dt} = \sum_{i+j=n}^{L} F_{ij}x^i y^j + H_2(x,y,t) \tag{2.36}$$

where H_1, H_2 are analytic in x and y, periodic in t with period 1, having all partial derivatives /t continuous, and $\lim_{r\to 0} H_1 r^{-L} = \lim_{r\to 0} H_2 r^{-L} = 0$ uniformily with respect to t, L an arbitrarily high integer such that if $(x_n, y_n) \varepsilon D$ (a sufficiently small neighbourhood of the origin) and if $x=p(t,x_n,y_n)$, $y=q(t,x_n,y_n)$ are the solutions of (2.36) with initial conditions $x=x_n$, $y=y_n$ for $t=0$ the mapping defined by

$$x_{n+1} = p(1,x_n,y_n) \quad , \quad y_{n+1} = q(1,x_n,y_n) \qquad (r=(x^2+y^2)^{1/2})$$

is the same as (2.28). In this sense we say that (2.36) generates (2.28). The proof of this theorem is given in (Ref. 15).

From this, it follows that the analysis of the stability of the fixed point $x_n = y_n = 0$ of (2.28) can be reduced to the analysis of the stability of the trivial solution of (2.36). This reduction is applied hereafter in order to solve the stability problem whenever the stability is defined by a finite number of terms of (2.28).

The stability of the trivial solution of (2.36) is examined here by two methods

a) use of a Lyapunov function (Ref. 28,31)

By performing on (2.36) a non linear change of variables $(x,y)\to(u,v)$ we are able to reduce it to the form

$$\frac{du}{dt} = v + \sum_{i=m}^{L} a_i u^i + \ldots$$
$$\frac{dv}{dt} = \sum_{i=n}^{L} b_i u^i + cv^n + \ldots \tag{2.37}$$

where c is an arbitrary constant and the unwritten terms are of higher degree than L. By means of simple calculations the following relations are found

(i) if m=n $\quad a_n = C_{n0} - D_{n0} \quad , \quad b_n = D_{n0}$

(ii) if m<n $\quad a_m = C_{m0} \quad , \quad b_n = D_{n0}$

(obviously if m>n, $a_n = -D_{n0}$, $b_n = D_{n0}$)

Some cases are solved by using the Lyapunov function

$$V(u,v) = \frac{v^2}{2} - \sum_{i=n}^{L} b_i \frac{u^{i+1}}{i+1} \tag{2.38}$$

which is positive definite if n is odd and $b_n = D_{n0} < 0$, and has the time derivative (with respect to (2.37)) given by

$$\frac{dV}{dt} = cv^{n+1} - b_n a_m u^{m+n} + \ldots$$

Keeping in mind that c is arbitrary, the following criteria may be readily obtained with n and m odd

Asymptotic stability	$m = n$	, $C_{n0} < D_{n0} < 0$
	$m < n$	, $D_{n0} < 0$, $C_{n0} < 0$
Instability	$m = n$	, $D_{n0} < 0$, $D_{n0} < C_{n0}$
	$m < n$	, $D_{n0} < 0$, $C_{n0} > 0$
	$m > n$	, $D_{n0} < 0$

b) "Pseudo-polar" coordinates

Here we slightly extend the method of Lyapunov (28) who treated the stability problem in relation to autonomous systems of the type (2.36) when, $H_1 \equiv H_2 \equiv 0$. In order to deal with the other remaining cases we use as in (28) the transcendental functions

$$\frac{dCs(w)}{dw} = - S_n(w) \ , \qquad \frac{dS_n(w)}{dw} = Cs^{2\ell-1}(w) \quad \text{with } Cs(0) = 1, \ S_n(0) = 0$$

The following equality is satisfied

$$Cs^{2\ell}(w) + S_n^2(w) = 1$$

Performing the change of variables $(x,y \rightarrow r,w)$: $x = rCs(w)$, $y = -r^{\ell}S_n(w)$ (2.36) is transformed into

$$\begin{aligned} \frac{dr}{dt} &= r^{\ell+1}[R(r,w) + E_r(r,w,t)] \\ \frac{dw}{dt} &= r^{\ell-1}[1 + rQ(r,w) + rE_w(r,w,t)] \end{aligned} \tag{2.39}$$

assuming (without loss of generality) that $F_{n0} = -1$. The functions E_r and E_w are periodic in t and w of class C^∞ and satisfy

$$\lim_{r\to 0} |E_r| \ r^{-(L-2\ell)} = \lim_{r\to 0} |E_w| \ r^{-(L-2\ell)} = 0$$

In the case of autonomous systems studied by Lyapunov (28) E_r and E_w were identically zero. The corresponding stability problem was reduced to the conditional stability of the solution $r \equiv 0$ of

$$\frac{dr}{dw} = \frac{r^2 R}{1+rQ} \tag{2.40}$$

under the condition $r \geqq 0$. This last problem was solved by the classical method of finding the solutions near r=0, as series expansions of the form

$$r(\theta) = r_0 + u_2(w) r_0^2 + u_3(w) r_0^3 + \ldots \tag{2.41}$$

r_0 being an arbitrary (sufficiently small) positive constant. By setting $u_2(0) = u_3(0) = \ldots = 0$, the sequence

$$u_2, u_3, u_4, \ldots \tag{2.42}$$

can be found uniquely from (2.41) and (2.40). If an integer $N \geqq 2$ exists such that u_N is the first non-periodic function of w in the sequence (2.42) then u_N is of the form

$$u_N = gw + v(w) \qquad (g \neq 0,\ v \text{ periodic in } w)$$

If g>0 the solution $r \equiv 0$ is unstable, if g<0 it is asymptotically stable.

In (15) it is shown that if $L \geqq N+2\ell-2$ then the trivial solution of (2.35) is also unstable if g>0 and asymptotically stable if g<0. The same can be said of the stability of the fixed point x=y=0 of the mapping (2.26) or (2.28).

II.5. Some Cases Where The Number Of Critical Multipliers is Higher Than 2

This section is devoted to the study of critical cases which can be reduced to the study of the critical case with two multipliers equal to 1, with a linear diagonal part.

In section II.4.1 for the complex multipliers case, when $w = 2k\pi/q$; k,q integers it appeared that in some cases a Lyapunov function could not be readily constructed because of the appearance of null denominator in the elimination procedure. Taking the qth power of the transformation, enables one to deal with a mapping with critical real multipliers. In chapter IV, this case is treated as an application of the stability analysis of periodic differential equations with parametric resonance.

The next case is that of a fourth order point mapping with critical complex multipliers, which can be considered as the non linear coupling of two mappings of type (2.13) and which can serve as a model for two dimensional oscillators (Ref.32). For sake of brevity, it is assumed that a linear complex change of variables has been performed on each of the two mappings in order to obtain on each the linear diagonal form (2.14)

$$\begin{aligned} R_i &= \exp[jw_i] r_i + \sum_{m=2}^{\infty} D_i^{(m)}(r_1, s_1, r_2, s_2) \\ S_i &= \exp[-jw_i] s_i + \sum_{m=2}^{\infty} D_i^{*(m)}(r_1, s_1, r_2, s_2) \end{aligned} \qquad i=1,2 \tag{2.43}$$

where $D_i^{(m)}$ is an homogeneous polynomial of degree m, D^* means conjugate of D. Before starting the stability study of the double point origin of (2.43), it is necessary to perform a non linear change of variables, to bring out the essential terms. This change must be non singular (non vanishing Jacobian at the origin) to ensure that the resulting point mapping has the same stability property as the original (2.43). This condition is clearly satisfied by :

$$\begin{aligned} r_i &= x_i + \sum_{j=2}^{N} X_i^{(j)}(x_1,y_1,x_2,y_2) \\ & \qquad\qquad\qquad\qquad i=1,2 \\ s_i &= y_i + \sum_{j=2}^{N} Y_i^{(j)}(x_1,y_1,x_2,y_2) \end{aligned} \tag{2.44}$$

N arbitrarily high integer and $X_i^{(j)}$, $Y_i^{(j)}$ are homogeneous polynomials of degree j. The object of this change is to cancel the maximum of terms in the resulting transformation, which will be written in terms of the x_i, y_i, i=1,2 variables :

$$\begin{aligned} X_i &= \exp[jw_i]x_i + \sum_{j=2}^{\infty} C_i^{(j)}(x_1,y_1,x_2,y_2) \\ Y_i &= \exp[-jw_i]y_i + \sum_{j=2}^{\infty} C_i^{*(j)}(x_1,y_1,x_2,y_2) \end{aligned} \tag{2.45}$$

Putting (2.44) in (2.43), identifying with (2.45), the elimination of terms of degree less than m (<N), leads to a set of equations of the form :

$$\begin{aligned} & X_i^{(m)}(\exp(jw_1)x_1, \exp(-jw_1)y_1, \exp(jw_2)x_2, \exp(-jw_2)y_2)+\ldots= \\ & \qquad \exp[jw_i]X_i^{(m)}(x_1,y_1,x_2,y_2)+ D_i^{m}(x_1,y_1,x_2,y_2)+\ldots \end{aligned} \tag{2.46}$$

where the dotted lines represent the m degree terms coming from $\sum_{j=2}^{m-1} X_i^{(j)}(X_1,Y_1,X_2,Y_2)$ and $\sum_{j=2}^{m-1} D_i^{(j)}(r_1,s_1,r_2,s_2)$.

It is easy to see that each of these forms introduces coefficients of less than m degree terms and thus coefficients that have already been determined in the preceding steps of the elimination of less than m degree terms. Let

$$\begin{aligned} D_i^{(m)}(r_1,s_1,r_2,s_2) &= D_i^{m_1 n_1 m_2 n_2} r_1^{m_1} s_1^{n_1} r_2^{m_2} s_2^{n_2} \\ & \qquad\qquad\qquad\qquad m_1+n_1+m_2+n_2=m \\ X_i^{(m)}(x_1,y_1,x_2,y_2) &= X_i^{m_1 n_1 m_2 n_2} x_1^{m_1} y_1^{n_1} x_2^{m_2} y_2^{n_2} \end{aligned}$$

Equations (2.46) can be written as a set of equations (linear with respect to the coefficients $X_i^{m_1 n_1 m_2 n_2}$, $D_i^{m_1 n_1 m_2 n_2}$) of the form

$$(\exp[j(m_1-n_1)w_1+j(m_2-n_2)w_2]- \exp[jw_i])\, X_i^{m_1 n_1 m_2 n_2} = D_i^{m_1 n_1 m_2 n_2} + R_i \tag{2.47}$$

R_i represents the contribution of the dotted lines in (2.46). Provided that the coefficient of $X_i^{m_1 n_1 m_2 n_2}$ is non null, it would be possible to fix this coefficient so as to cancel the term $(x_1^{m_1} y_1^{n_1} x_2^{m_2} y_2^{n_2})$ in (2.45). The only remaining terms are those such that

$$(m_1-n_1)w_1+(m_2-n_2)w_2- w_i = 2k\pi \;,\; i=1,2,\; k \text{ integer} \tag{2.48}$$

Whatever the values of w_1 and w_2 there are some terms that can never be eliminated. Indeed, (2.48) holds whenever the integer coefficients of w_1 and w_2 in (2.48) are null, that is :

for X_1 (i=1) $m_1-n_1-1=0$, $m_2-n_2=0$

for X_2 (i=2) $m_1-n_1=0$, $m_2-n_2-1=0$

It is assumed in what follows that (2.48) is satisfied for these relations of m_1,n_1,m_2,n_2 only ; in other words, that there is no linear combination with integer coefficients of w_1 and w_2 which is multiple of 2π. Then (2.45) is written

$$X_1=\exp[jw_1]x_1+\sum_{i+j=1}^{i+j=\frac{N-1}{2}} C_1^{i+1,i,j,j}\, x_1^{i+1} y_1^{i}\, x_2^{j}\, y_2^{j} \quad ; \; Y_1= X_1^{*}$$

$$X_2= \exp[jw_2]x_2+\sum_{i+j=1}^{i+j=\frac{N-1}{2}} C_2^{i,i,j+1,j}\, x_1^{i}\, y_1^{i}\, x_2^{j+1} y_2^{j} \; ; \; Y_2= X_2^{*} \tag{2.49}$$

In polar coordinates $X_i= R_i\exp(jV_i)$, $x_i= r_i\exp(jv_i)$, i=1,2 ; (2.49) is written

$$R_1^2= r_1^2+ F(r_1^2,r_2^2)\ ,\ V_1= v_1$$

$$R_2^2= r_2^2+ G(r_1^2,r_2^2)\ ,\ V_2= v_2 \tag{2.50}$$

which, obviously, reduces to a second order point mapping in the r variables with the critical case : multipliers equal to 1, linear part. If there are third degree terms in (2.49), then (Ref. 33)

$$R_1^2= r_1^2+ A_1r_1^2+ B_1r_1^2r_2^2$$

$$R_2^2= r_2^2+ A_2r_2^2+ B_2r_1^2r_2^2$$

with $A_1= 2\ \mathrm{Re}(C_1^{2100}\exp[-jw_1])$, $B_1= 2\ \mathrm{Re}(C_1^{1011}\exp[-jw_1])$, $A_2= 2\ \mathrm{Re}(C_2^{0021}\exp[-jw_2])$, $B_2= 2\ \mathrm{Re}(C_2^{1110}\exp[-jw_2])$ where Re[.] means real part of [.] . With the notation introduced in section II.4.2.

$$Q(r_1^2,r_2^2)= r_1^2r_2^2([B_2-A_1]r_1^2+[A_2-B_1]r_2^2)$$

$$P(r_1^2,r_2^2)= A_1r_1^6+B_1r_1^4r_2^2+B_2r_1^2r_2^4+A_2r_2^6$$

the invariant directions (Q=0) are :

$$r_1^2= 0,\ r_2^2= 0 \quad \text{and} \quad (B_2-A_1)r_1^2+ (A_2-B_1)\, r_2^2= 0 \ \text{if}\ \frac{A_2-B_1}{A_1-B_2} >0$$

The stability conditions are obtained by writting P<0 on each invariant direction giving

$$A_1<0,\ A_2<0 \quad \text{and if}\ \frac{A_2-B_1}{A_1-B_2} >0 \qquad \frac{A_1A_2-B_1B_2}{A_1-B_2} <0$$

Similarly, it is relatively easy to show that the following critical case

$$
\begin{aligned}
x_{n+1} &= x_n + X(x_n,u_n,v_n) \\
u_{n+1} &= \exp[jw]u_n + U(x_n,u_n,v_n) \\
v_{n+1} &= \exp[-jw]v_n + U^*(x_n,u_n,v_n)
\end{aligned} \tag{2.51}
$$

that is a critical case with one multiplier equal to one and two complex multipliers, can be reduced to the study of a second order point mapping with two multipliers equal to one. Malkin (12) did this for differential equation and the transposition to the discrete case is straightforward.

II.6. Conclusion

At first sight it may appear that this study of critical cases is essentially mathematical and thus from a practical point of view purely academic. This view would be justified if the notion of critical case were not extended. It is true that for a real physical system, a critical case corresponds, in the space of the physical parameters, to very particular situations. However, if the notion of bifurcation is added then the interest of studying critical cases become obvious, as has been made clear by several authors (Ref.2,16,34)

Bifurcation, i.e., a qualitative change in the behaviour of a system is generally linked to the occurence of a critical case and may be of practical interest. When considering a physical system working in the vicinity of a critical case (in the parameter space) it is of interest to know the type of bifurcation which may occur with the variation of certain parameters in order to predict whether such an occurence presents any risks of irreversibility. These considerations led to the definition of dangerous and non dangerous stability boundaries (Bautin, 11). They correspond to two types of bifurcation:

- if the critical case is stable, corresponding to the non dangerous boundary, the qualitative change in the behaviour takes place in a reversible way, the effects on the solutions being small in relation to the perturbation of the parameters

- if the critical case is unstable, the situation may be drastically different and even small parameter perturbations may cause significant modification of the behaviour, and even, irreversibility. The boundary is called dangerous.

Except for (i) the simplest critical case of one multiplier unity
(ii) the complex multiplier case (also called Hopf bifurcation by mathematicians) there is no general approach to the study of bifurcation. This is due both to the wide variety of critical cases which are found, and to the fact that bifurcation manifests itself differently in each case.

APPENDIX 1 - Calculation of G_4

Consider the mapping

$$r_{n+1} = r_n \cos w + s_n \sin w + \sum_{m+n=2}^{3} a_{mn} r^m s^n + \dots$$
$$s_{n+1} = r_n \sin w + s_n \cos w + \sum_{m+n=2}^{3} b_{mn} r^m s^n + \dots \qquad (2.52)$$

on which is carried out a linear change with complex variables $x=r+js$, $y=r-js$. Then

$$x_{n+1} = x_n \exp[jw] + \sum_{m+n=2}^{3} C_{mn} x^m y^n + \dots$$
$$y_{n+1} = y_n \exp[-jw] + \sum_{m+n=2}^{3} C^*_{mn} x^n y^m + \dots \qquad (2.53)$$

and

$$C_{20} = \frac{1}{4}[a_{20}+b_{11}-a_{02}+j(b_{20}-a_{11}-b_{02})]$$
$$C_{11} = \frac{1}{2}[(a_{20}+a_{02})+j(b_{20}+b_{02})]$$
$$C_{02} = \frac{1}{4}[(a_{20}-b_{11}-a_{02})+j(b_{20}+a_{11}-b_{02})]$$
$$C_{30} = \frac{1}{8}[(a_{30}+b_{21}-a_{12}-b_{03})+j(b_{30}-a_{21}-b_{12}+a_{03})] \qquad (2.54)$$
$$C_{21} = \frac{1}{8}[(3a_{30}+b_{21}+a_{12}+3b_{03})+j(3b_{30}-a_{21}+b_{12}-3a_{03})]$$
$$C_{12} = \frac{1}{8}[(3a_{30}-b_{21}+a_{12}-3b_{03})+j(3b_{30}+a_{21}+b_{12}+3a_{03})]$$
$$C_{03} = \frac{1}{8}[(a_{30}-b_{21}-a_{12}+b_{03})+j(b_{30}+a_{21}-b_{12}-a_{03})]$$

The Lyapunov function chosen is in the form

$$V(x,y) = xy + \sum_{m+n=3}^{4} v_{mn} x^m y^n + \dots$$

whose increment $\Delta V = V(x_{n+1}, y_{n+1}) - V(x_n, y_n)$ can be expressed

$$\Delta V = G_4 x_n^2 y_n^2 + \dots$$

by a proper choice of the v_{mn} coefficients.

$$\Delta V=(x_n\exp[jw]+\sum_{k+\ell=2}^{3} C_{k\ell}x_n^k y_n^\ell)(y_n\exp[-jw]+\sum_{k+\ell=2}^{3} \overset{::}{C}_{k\ell}x_n^\ell y_n^k)$$

$$+\sum_{m+n=3} v_{mn}(x_n\exp[jw]+\sum_{k+\ell=2} C_{k\ell}x_n^k y_n^\ell)^m(y_n\exp[-jw]+\sum_{k+\ell=2} \overset{::}{C}_{k\ell}x_n^\ell y_n^k)^n$$

$$+\sum_{k+\ell=4} v_{k\ell}x_n^k y_n^\ell \exp[j(k-\ell)w]-x_n y_n-\sum_{k+\ell=3}^{4} v_{k\ell}x_n^k y_n^\ell+\ldots$$

The dotted lines represent terms of degree higher than 4. Elimination of third degree terms lead to

$$\begin{aligned}
&\overset{::}{C}_{02}\exp[jw]+v_{30}(\exp(3jw)-1)=0\\
&C_{20}\exp[-jw]+\overset{::}{C}_{11}\exp[jw]+v_{21}(\exp(jw)-1)=0\\
&C_{11}\exp[-jw]+\overset{::}{C}_{20}\exp[jw]+v_{12}(\exp(-jw)-1)=0\\
&C_{02}\exp[-jw]+v_{03}(\exp(-3jw)-1)=0
\end{aligned}\qquad(2.56)$$

This gives the v_{mn} $(m+n=3)$ coefficients provided that $w\neq 2k\pi/3$ or 2π. Cancellation of the fourth order terms is made in the same way and provided $w\neq 2k\pi/4$ all these terms can be cancelled except the terms $G_4x_n^2y_n^2$. It is easy to see that the second step of the elimination does not affect the coefficient G_4. This can be obtained from equations (2.55) and (2.56); one gets :

$$G_4=\overset{::}{C}_{21}\exp[jw]+C_{21}\exp[-jw]+C_{20}\overset{::}{C}_{20}+C_{11}\overset{::}{C}_{11}+C_{02}\overset{::}{C}_{02}+3v_{30}C_{02}\exp[2jw]$$

$$+v_{21}(\overset{::}{C}_{20}\exp[2jw]+2C_{11})+v_{12}(C_{20}\exp[-2jw]+2\overset{::}{C}_{11})+3v_{03}\overset{::}{C}_{20}\exp[-2jw]$$

and after some calculations a (perhaps) simpler form

$$G_4=\overset{::}{C}_{21}\exp[jw]+C_{21}\exp[-jw]-2C_{02}\overset{::}{C}_{02}-C_{11}\overset{::}{C}_{11}+\overset{::}{C}_{20}\overset{::}{C}_{11}[\exp[3jw](1-\exp(jw))^{-1}$$

$$+2\exp[jw](1-\exp(-jw))^{-1}]+C_{20}C_{11}[2\exp[-jw](1-\exp(jw))^{-1}+\exp(-3jw)$$

$$(1-\exp(-jw))^{-1}]$$

APPENDIX 2

Consider the mapping: $x_{n+1}=x_n + \sum_{j=m}^{\infty} f_j(x_n,y_n)$; $y_{n+1}=y_n + \sum_{j=m}^{\infty} g_j(x_n,y_n)$

f_j, g_j being homogeneous polynomials in x_n, y_n, m integer ≥ 2. In section II.4.2, this critical case was studied for the case when the m degree approximation is sufficient to determine stability or instability. In what follows some partial answers are given for cases when such an approximation is not sufficient. By the polar coordinate change $x = r \cos w$, $y = r \sin w$ (2.57) is written

$$\begin{aligned} r_{n+1}^2 &= r_n^2 + P_m(w)r_n^{m+1} + P_{m+1}(w)r_n^{m+2} + \dots \\ w_{n+1} &= w_n + Q_m(w)r_n^{m-1} + Q_{m+1}(w)r_n^{m} + \dots \end{aligned} \tag{2.58}$$

where

$$P_m(w) = 2[\cos w\, f_m(\cos w,\sin w) + \sin w\, g_m(\cos w,\sin w)]$$

$$Q_m(w) = \cos w\, g_m(\cos w,\sin w) - \sin w\, f_m(\cos w,\sin w)$$

P_{m+1}, Q_{m+1} are easily obtainable from P_m, Q_m by substituting f_{m+1}, g_{m+1} for f_m, g_m in the case m>2. For m=2 some additional terms are needed.

1st case : $Q_m(w) \neq 0, \forall w$

$h_0 = \frac{1}{2}\int_0^{2\pi} \frac{P_m(w)}{Q_m(w)}\, dw = 0$. By the following change of variables

$R = r \exp\left[\int_0^w \frac{P_m(w)}{2Q_m(w)}\, dw\right] = r\, e(w)$, (2.58) is written

$$\begin{aligned} R_{n+1}^2 &= R_n^2 + \bar{P}_{m+1}(w)\, R^{m+2} + \dots \\ w_{n+1} &= w_n + \bar{Q}_m(w)\, R^{m-1} + \dots \end{aligned} \tag{2.59}$$

where

$$\bar{P}_{m+1}(w) = (P_{m+1}Q_m - P_m Q_{m+1})Q_m^{-1} e(w)^{-m}$$

$$\bar{Q}_m(w) = Q_m e(w)^{-m+1} \qquad (\neq 0, \forall w)$$

The stability problem can be solved by using the following Lyapunov function

$$V = R \exp\left[R\int_0^w h(u)du\right]$$

with h(u) as a periodic function of period 2π in u and of zero mean value.

Then

$$\Delta V = R^{m+1} V_n \left[\frac{\bar{P}_{m+1}}{2} + h(w_n)\bar{Q}_m + R\ldots\right] = R^{m+1} V_n (h_1 \bar{Q}_m + R\ldots)$$

if $\quad h(w) = h_1 - \dfrac{\bar{P}_{m+1}}{2\bar{Q}_m} \quad$ with $\quad h_1 = \dfrac{1}{2\pi}\displaystyle\int_0^{2\pi} \frac{\bar{P}_{m+1}(w)}{2\bar{Q}_m(w)}\, dw$

The fixed point origin is asymptotically stable (unstable) if

$$hQ_m < 0 \quad (hQ_m > 0) \quad , \text{ since sign } \bar{Q}_m = \text{ sign } Q_m$$

When $h_1 = 0$, higher degree terms are needed and the above procedure can be continued (change of variables, choice of a Lyapunov function) until the first $h_i \neq 0$. In general, for

$$R_{n+1}^2 = R_n^2 + \tilde{P}_{m+k} R^{m+k+1} + \ldots$$

$$w_{n+1} = w_n + \tilde{Q}_m R^{m-1} + \ldots$$

using the Lyapunov function $V = R \exp[R^k \int_0^w h(u)du]$ where $h(u)$ is a 2π periodic function with zero mean value, the fixed point origin is stable (unstable) if

$$h_k Q_m < 0 \quad (>0) \quad ; \quad h_k = \frac{1}{2\pi}\int_0^{2\pi} \frac{\tilde{P}_{m+k}}{2\tilde{Q}_m}\, dw$$

2nd case. Invariant directions exist.

Because of the numerous subcases (36-37) what follows is restricted to those which seem the most important : $Q_m(w) = 0$ has roots of multiplicity one. First there is assumed to be a single root, i.e.

$$Q_m(w) = (\sin w - p_0 \cos w) Q_{m-1}(w)$$

$$P_m(w) = (\sin w - p_0 \cos w) P_{m-1}(w) \qquad Q_{m-1}(w) \neq 0$$

Using a Lyapunov function $V = r \exp\left[\int_0^w h(u)du\right]$ (h periodic, zero mean value, to be defined) we get

$$\Delta V = r^m V_n \left[(\sin w - p_0 \cos w)\left(\frac{P_{m-1}}{2} + Q_{m-1} h\right) + r\left(\frac{P_{m+1}}{2} + Q_{m+1} h\right) + \ldots\right] \tag{2.60}$$

It is possible to determine a function $g(w)$, 2π periodic, sign constant and

$$\frac{P_{m-1}}{2} + Q_{m-1} h(w) = g(w)(\sin w - p_0 \cos w)$$

$$\int_0^{2\pi} \frac{g(w)(\sin w - p_0 \cos w)}{Q_{m-1}}\, dw = \int_0^{2\pi} \frac{P_{m-1}}{2Q_{m-1}}\, dw$$

Then (2.60) becomes

$$\Delta V = r^m V_n \left[(\sin w_n - p_0 \cos w_n)^2 g(w_n) + r\left(\frac{P_{m+1}Q_{m+1} - Q_{m+1}P_{m-1}}{2Q_{m-1}} + \frac{Q_{m+1}}{Q_{m-1}}(\sin w - p_0 \cos w) g(w)\right.\right.$$

Chosing $\operatorname{sign}[g(w)]=\operatorname{sign}\left[\frac{P_{m+1}Q_{m-1}-Q_{m+1}P_{m-1}}{Q_{m-1}}\right]_{w=w_0}$, it can be stated that

0 stable (unstable) if $\left[\frac{P_{m+1}(w)Q_{m-1}(w)-Q_{m+1}(w)P_{m-1}(w)}{Q_{m-1}(w)}\right]_{w=w_0} < 0 \ (>0)$

In the case when the equation $Q(w)=0$ has several roots, the demonstration is slightly more difficult but uses the same idea in defining a function $h(w)$ and $g(w)$. For further details see (Ref. 35) for differential equations and (Ref. 38) for recurrent equations.

The stability is assured if on each invariant direction w_i where $P_m(w_i)=C$ the quantity $g_i= \left[\frac{P_{m+1}Q_{m-1}-Q_{m+1}P_{m-1}}{Q_{m-1}}\right]_{w=w_i}$ is negative.

If there is one direction w_i on which $g_i>0$ the fixed point origin is unstable.

CHAPTER III
DIFFERENTIAL SYSTEMS WITH PERIODIC COEFFICIENTS

In the last chapter, an example is given of associated point mapping to a system which can be described directly by non linear difference equation (electric power rectifiers). This chapter is devoted to a less straightforward application of associated point mapping : namely, non linear differential equations with periodic coefficients. Although it is known since Poincaré that it is possible to associate point mapping with such equation relatively few applications have been made using this method. The essential reason is that generally it is a difficult task to determine the analytical expressions of the associated mapping and there are few examples (Ref.41-42) where these expressions can be expressed by means of classical functions of analysis. In the following sections a way of calculating the first degree terms of the series expansions of the mapping associated with periodic coefficients is given. This will make it possible to carry out a study of the local properties of the equilibrium points (Ref.39) and the periodic solutions and to deal with the problem of parametric resonance (chapter IV). The method uses some of the work of Birkhoff (Ref. 30), developped by Lewis (Ref. 40). A summary of them is given in the following section. At this stage it is worth mentionning an other method which can be applied to periodic differential equations and which have some features in common with the method described in this chapter : the "stroboscopic method" (Minorski, 70)

III.1. Point Mapping And The Associated Formal Differential System

III.1.1. Generalities

For the sake of clarity and conciseness let us restrict ourselves in the first instance to the following particular point mapping τ :

$$x_1 = sx + \sum_{m,n} C_{mn} x^m y^n \qquad m+n=2,3\ldots \tag{3.1}$$
$$y_1 = s^{-1} y + \sum_{m,n} D_{mn} x^m y^n$$

which is analytic in the neighbourhood of the origin. The k-th iterate (k-th consequent of point x,y) defined when k is a positive integer can be expressed as a convergent series, defining the mapping τ^k :

$$x_k = s^k x + \sum_{m,n} C_{mn}^{(k)} x^m y^n \qquad m+n=2,3\ldots \tag{3.2}$$
$$y_k = s^{-k} y + \sum_{m,n} D_{mn}^{(k)} x^m y^n$$

where the $C_{mn}^{(k)}$, $D_{mn}^{(k)}$ coefficients are polynomials in s^k, s^{-k} and k of degree not greater than m+n. Furthermore the relation $\tau^{k+i} = \tau^k . \tau^i$ holds for all integral values of k and i . Let us now assume that it is possible to define the series x_k, y_k for real values of k so that the above relation also holds for all values of k and i (real) in a purely formal sense. Birkhoff showed that the series x_k, y_k satisfy the formal differential system

$$\frac{dx_k}{dk} = U(x_k, y_k) \quad ; \quad \frac{dy_k}{dk} = V(x_k, y_k)$$

where

$$U(x,y) = \left[\frac{\partial x_k}{\partial k}\right]_{k=0} \qquad V(x,y) = \left[\frac{\partial y_k}{\partial k}\right]_{k=0} \tag{3.3}$$

with initial conditions k=0, $x_0 = x$, $y_0 = y$. Inversely, the series x_k, y_k are determined uniquely by these equations and initial conditions. This fact will be used later for determining the mapping associated with a non linear differential equation.

Although this is not essential for our purpose, it may be useful here to say a few words about the fact that differential equation (3.3) must be considered as formal.

Indeed, if we consider a conservative point mapping eg, area preserving point mapping, the existence of a true differential system (3.3), that is with right hand sides U(x,y), V(x,y) convergent, would imply that there is an analytic first integral and consequently that the mapping is completely integrable. This, obviously, is not generally the case. Many examples of mappings which are not completely integrable can be found in the literature (see chapter IV).

Even in the association of the formal differential system (3.3) to the mapping (3.1) some care must be taken. Lewis (40) showed, in his paper, "formal power series transformations", that the possibility of such a formal association depends on the values of the multipliers of the mapping which must give the so-called "pseudo-incommensurable" case. In general terms, a non "pseudo-incommensurable" case occurs when there is a linear combination with integer coefficients of a particular determination of the logarithm of the eigen values giving a non-zero multiple of $2\pi j$. For instance, in the case of a second order point mapping like (3.1) a non "pseudo-incommensurable" case occurs when the argument of the multiplier s is equal to $\frac{2k\pi}{q}$, whatever k and q integers :

$$(s_1 = r\exp[jw],\ s_2 = r\exp[-jw] \ ,\quad n_1 \mathrm{Log} s_1 + n_2 \mathrm{Log} s_2 = (n_1+n_2)\mathrm{Log} r + j(n_1-n_2)w$$
$$= 2jk\pi \quad \text{if and only if} \quad n_1 = -n_2 \quad)$$

These cases can be solved by taking an integral power of τ. In the preceeding example for instance when $\frac{2k\pi}{q}$, τ^q gives a pseudo-incommensurable case.

All this points to the fact that the behaviour of the solutions of autonomous point mappings presents more variety than those of differential equations of the same order. For instance, in differential equations there is no equivalent to a type 2 or 3 saddle. In fact, when these singularities are present the mapping gives a non pseudo-incommensurable case with negative multipliers (argument $-\pi$). Consequently, care must be taken when trying to apply the method developed for differential equations to recurrent equations : in particular since recurrent equations can generally be associated to differential equations of greater order (Poincaré's surface of section method). These ideas are not developed further since in what follows the reverse case is considered namely, given a differential system of the form (3.3), find the mapping (3.1) from which it originates.

Before going on, it is necessary to describe more precisely the form of the system (3.3) and in order to do that we shall try to give the expressions of the series x_k, y_k, defining the mapping τ^k.

III.1.2. Determination Of The Series x_k, y_k

This can be done by writing $\tau^{k+1} = \tau . \tau^k$. That is

$$x_{k+1} = s^{k+1} x + \sum_{m,n} C_{mn}^{(k+1)} x^m y^n$$

$$= s x_k + \sum_{m,n} C_{mn} x_k^m y_k^n \qquad m+n \geq 2 \qquad (3.4)$$

$$y_{k+1} = s^{-(k+1)} y + \sum_{m,n} D_{mn}^{(k+1)} x^m y^n$$

$$= s^{-1} y_k + \sum_{m,n} D_{mn} x_k^m y_k^n$$

With (3.2) and (3.4), identifying the coefficients of terms of same degree with respect to x and y, one gets a system of difference equations from which the $C_{mn}^{(k)}$ and $D_{mn}^{(k)}$ coefficients can be calculated. For the sake of conciseness, we will restrict ourselves to give some details of the calculations of the C_{20}, C_{30} coefficients only, the others being given in the appendix.

Equating in (3.4) the coefficients of the terms in x^2 and x^3, one gets the following difference equations :

$$C_{20}^{(k+1)} = sC_{20}^{(k)} + s^{2k} C_{20}$$

$$C_{30}^{(k+1)} = sC_{30}^{(k)} + s^{3k} C_{30} + 2C_{20} s^k C_{20}^{(k)} + C_{11} s^k D_{20}^{(k)} \qquad (3.5)$$

These equations are linear difference equations with respect to the $C_{mn}^{(k)}$, $D_{mn}^{(k)}$ coefficients, and are of the form :

$$C_{mn}^{(k+1)} = sC_{mn}^{(k)} + s^{m-n} C_{mn} + P_{mn}$$

where P_{mn} involves the coefficients $C_{ij}^{(k)}$, $D_{ij}^{(k)}$ with 2 $i+j<m+n$, i.e., coefficients that have been determined by former difference equations. Of course, the number of terms included in P_{mn} increases with the order m+n, and does the complexity of the calculations. In practical terms, it is reasonable to undertake these calculations for the degrees N and N+1, N being the minimum degree of the non linear part of (3.1). However, it will be shown in the next chapter, that generally these terms enable one to solve some important problems such as stability in the critical case, the study of parametric resonance, ...

From (3.5) we get

$$C_{20}^{(k)} \doteq C_{20} \frac{s^{2k} - s^k}{s^2 - s} \qquad (3.7)$$

$$C_{30}^{(k)} = \left(C_{30} - \frac{2C_{20}^2}{s - s^2} - \frac{C_{11} D_{20}}{s^{-1} - s^2}\right) \frac{s^{3k} - s^k}{s^3 - s} + \frac{2C_{20}^2}{s - s^2} \cdot \frac{s^{2k} - s^k}{s^2 - s} + \frac{C_{11} D_{20}}{s^{-1} - s^2} \frac{1 - s^k}{1 - s}$$

III.1.3. Formal Differential System

Let us consider s complex, s=exp[jw], this particular form of the multiplier is chosen in order that the calculations in this section could be used in chapter IV. (3.1) is obtained from a real mapping by a linear complex change of variables. This implies $D_{mn} = C^{*}_{mn}$, where C^{*}_{mn} denotes conjugate of C_{mn}.

From (3.2) with the $C_{mn}^{(k)}$ given in the appendix, one gets the following formal differential system :

$$\frac{dx_k}{dk} = \mathrm{Log}(s)x_k + \sum_{m,n} \left[\frac{d}{dk} C_{mn}^{(k)}\right]_{k=0} x_k^m y_k^n$$

$$\frac{dy_k}{dk} = -\mathrm{Log}(s)y_k + \sum_{m,n} \left[\frac{d}{dk} D_{mn}^{(k)}\right]_{k=0} x_k^m y_k^n \tag{3.8}$$

That is for s=exp[jw]

$$\frac{dx_k}{dk} = jw\, x_k + \sum_{m,n} \left[\frac{d}{dk} C_{mn}^{(k)}\right]_{k=0} x_k^m y_k^n$$

$$\frac{dy_k}{dk} = -jw\, y_k + \sum_{m,n} \left[\frac{d}{dk} C_{mn}^{*(k)}\right]_{k=0} x_k^m y_k^n \tag{3.9}$$

III.2. Determination Of The Mapping Associated With A Non Linear Differential Equation With Periodic Coefficients

Let us consider the non linear differential equation of the form

$$\frac{d^2x}{dt^2} + a^2(t)x = f(x, \frac{dx}{dt}, t) \tag{3.10}$$

where $a^2(t)$, $f(x, \frac{dx}{dt}, t)$ are periodic with period T with respect to the independent time variable and $f(x, \frac{dx}{dt}, t)$ has a series development which begins with terms of at least 2nd degree with respect to the x, $\frac{dx}{dt}$ variables :

$$f(x,y,t) = \sum_{m,n} A_{mn}(t)\, x^m y^n \qquad m+n \geq 2 \tag{3.11}$$

III.2.1. Brief Description Of The Method (Ref.33)

We will first assume that the periodic coefficients a(t) and $A_{mn}(t)$ of (3.10) and (3.11) are piecewise constant on L subintervals S_i of the period T, that is :

$$a(t) = a_i \ , \ A_{mn}(t) = A_{mn}^{\ i} \quad \text{for } t \in S_i$$

$$S_i = \left\{ t;\ nT+t_i \leq t < nT+t_{i+1} \right\};\ i=0,1,\ldots L-1$$

We need in addition a further assumption (one which is physically meaningful and thus not very restrictive practically) : there is no discontinuity in the state variables $(x, y = \frac{dx}{dt})$ of (3.10) at the instant $t=nT+t_i$, that is at the instants when there is a discontinuity in the coefficients of the differential equations (3.10).

With these assumptions the principle of the method of determining the mapping associated with (3.10) becomes clear :

1°- on each subinterval S_i where the coefficients are constant, calculate the partial mapping τ_i by identifying the corresponding differential equation derived from (3.10) with the system (3.9) or more generally (3.3)

2°- As a consequence of continuity in the state variables, the mapping τ over one period T is the product of the partial ones

$$\tau = \tau_{L-1}.\tau_{L-2} \cdots \tau_1.\tau_0$$

As above, we shall not go into the details (see appendix) restricting ourselves to indicate the main calculations on the coefficients C_{20}, C_{30} only.

III.2.2. Some Calculations

By the linear complex change of variables $X=x+j\frac{y}{a_i}$, $Y=x-j\frac{y}{a_i}$ on the subinterval S_i, the differential equation (3.10) can be written as :

$$\frac{dX}{dt} = -ja_iX+jF_i(X,Y)/a_i$$
$$\frac{dY}{dt} = ja_iY-jF_i(X,Y)/a_i \qquad (3.12)$$

where
$$F_i(X,Y)=f_i(\frac{X+Y}{2}, \frac{X-Y}{2j} a_i)$$

Before identifying (3.12) and (3.9) it is necessary to make a change of time scale in one of the two systems (3.12) or (3.9). Indeed, what is required is for the mapping to link the states of the system over the interval S_i :

$$x_0= X(nT+t_i) \;;\; y_0= Y(nT+t_i) \quad \text{and} \quad x_1= X(nT+t_{i+1}) \;;\; y_1= Y(nT+t_{i+1}) \qquad (3.13)$$

For this it is sufficient to write :

$t = nT+t_i+k(t_{i+1}-t_i) = nT+t_i+kh_i$; h_i the length of the S_i subinterval.

After this change of variables (3.12) is written (for convenience) in the form

$$\frac{dX}{dk} = -ja_ih_iX + ja_ih_i \sum_{m,n} b_{mn}X^mY^n$$
$$\frac{dY}{dk} = ja_ih_iY - ja_ih_i \sum_{m,n} b^*_{mn} X^mY^n \qquad (3.14)$$

with
$$\sum_{m,n} b_{mn}X^mY^n= F_i(X,Y)/a_i^2$$

The identification between (3.14) and (3.9) gives the coefficients of the mapping τ_i which is expressed in terms of the variables :

$X = x+jy/a_i$, $Y = x-jy/a_i$

First, for the linear terms, one gets : $w_i = -a_i h_i \Rightarrow s=\exp[-ja_i h_i]$ and for the C_{20} coefficient

$$\frac{d}{dk}\left[C_{20}\,\frac{s^{2k}-s^{k}}{s^{2}-s}\right]_{k=0} = ja_i h_i b_{20}$$

That is

$$C_{20}\,\frac{-ja_i h_i}{s^{2}-s} = ja_i h_i b_{20} \qquad \text{and} \qquad C_{20} = b_{20}(s-s^{2})$$

Analogous calculations are made for all the coefficients. The results can be found in the appendix.

Remarks : 1°) In section III.2.1 it was assumed that the coefficients were piecewise constant. This was a necessary assumption since the formal differential system (3.9) is an autonomous one, independent of the independent variable k. When the periodic coefficients a(t), $A_{mn}(t)$ are continuous functions of the independent time variable, the method can be applied by making a piecewise constant approximation of the coefficients for instance by dividing the period into n equal intervals :

$$h_i = \frac{T}{n}\ ,\ a(t) = a_i = a\left(\frac{iT}{n}\right)\ ;\ A_{mn}(t) = A_{mn}{}^{i} = A_{mn}\left(\frac{iT}{n}\right)\ \text{for}$$

$$t\varepsilon\left[\frac{iT}{n}\ ,\ (i+1)\,\frac{T}{n}\right],\ i=0,1,\ldots,n-1$$

This leads to the difficult mathematical problem of the functional continuity of the solution of the differential equation with respect to the periodic coefficients. However, except for some very particular instances, it is generally the case that the mapping τ^{*} tends toward the true mapping τ for $n\rightarrow\infty$. The reader is referred to the paper by Garrett Birkhoff "on product integration" (Ref. 77).

Of course, the determination is carried out numerically, and in (Ref. 33) a numerical program is given, with some applications of the method.

2°) Before obtaining τ by multiplication of the partial mappings τ_i one must take into account the fact that the τ_i are not expressed in terms of the same variables when a(t) is not constant, indeed

$$X=x+jy/a_i\ ,\quad Y=x-jy/a_i$$

So, before multiplications, all the partial mappings must be expressed in the same base.

III.3. Other Forms Of Differential Equations

Up to now, the calculations have been restricted to a particular form of differential equation : namely

$$\frac{d^2x}{dt^2} + a(t)x = f(x,\frac{dx}{dt},t)$$

However the method can be extended to other forms of differential system. In this section the case of second order differential equations is considered. The restriction to second order is not essential since the method can theoretically be applied to equations of any order. To find the form of differential equations to which the method can be applied we will start from recurrent equations in various forms and associate with them a formal differential system as in section 3.1 .

In critical cases in the sense of Lyapunov non linear terms are needed in order to determine stability. For this reason we will restrict ourselves to systems which have such critical cases. However, the method can also be applied to non critical equations.

III.3.1. Mapping With Complex Multipliers

As we have already seen, to a recurrent equation (in terms of complex variables) of the form

$$X_1 = sX + \sum_{m,n} C_{mn} X^m Y^n$$
$$Y_1 = s^{-1} Y + \sum_{m,n} C^{*}_{mn} X^n Y^m \qquad (3.15)$$

with s=exp[-jw] a formal differential system of the form is associated

$$\frac{dX}{dk} = -jwX + \sum_{m,n} \ldots$$
$$\frac{dY}{dk} = jwY + \sum_{m,n} \ldots \qquad (3.16)$$

Conversely, the recurrent equation (3.15) could be associated with any second order differential system which could be written in the form (3.16). This is the case for the differential equation

$$\frac{d^2x}{dt^2} + a^2(t)x = f(x, \frac{dx}{dt}, t) \qquad (3.17)$$

and the differential system

$$\frac{dx}{dt} = a(t)x + b(t)y + P(x,y,t)$$
$$\frac{dy}{dt} = c(t)x - a(t)y + Q(x,y,t) \qquad \text{with } a^2 + bc < 0 \qquad (3.18)$$

where f,P,Q are functions with periodic coefficients whose series developments begin with terms of at least second degree with respect to the x,y variables.

III.3.2. Mapping With Real Multipliers

Consider the mapping (real variables)

$$X_1 = X + \sum_{m,n} C_{mn} X^m Y^n$$
$$Y_1 = sY + \sum_{m,n} D_{mn} X^m Y^n \qquad m+n \geq 2, \ s>0 \qquad (3.19)$$

The x_k, y_k series are written

$$X_k = X + \sum_{m,n} C_{mn}^{(k)} X^m Y^n$$
$$Y_k = s^k Y + \sum_{m,n} D_{mn}^{(k)} X^m Y^n \tag{3.20}$$

and, thus the formal differential system

$$\frac{dX_k}{dk} = \sum_{m,n} a_{mn} X^m Y^n$$
$$\frac{dY_k}{dk} = \text{Log}(s) Y + \sum_{m,n} b_{mn} X^m Y^n \tag{3.21}$$

with

$$a_{mn} = \left[\frac{d}{dk} C_{mn}^{(k)}\right]_{k=0} \qquad b_{mn} = \left[\frac{d}{dk} D_{mn}^{(k)}\right]_{k=0}$$

The mapping (3.19) can be associated with any differential equation that can be reduced to the form (3.21), for example

$$\frac{d^2x}{dt^2} + \alpha \frac{dx}{dt} = f(x, \frac{dx}{dt}) \quad , \qquad \alpha > 0$$

expressed in terms of the variables $X = \alpha x + y$, $Y = y$

In the case where $s = 1$, that is in the critical case where both multipliers are equal to one, the mapping (3.19) can be associated with a differential system with no linear terms in the right side of equations

$$\frac{dx}{dt} = P(x,y)$$
$$\frac{dy}{dt} = Q(x,y) \tag{3.22}$$

Let us consider the case with two multipliers equal to 1 but where the matrix of the linear part is not diagonal

$$X_1 = X + \delta Y + \sum_{m,n} C_{mn} X^m Y^n$$
$$Y_1 = Y + \sum_{m,n} D_{mn} X^m Y^n \tag{3.23}$$

The X_k, Y_k series can be written

$$X_k = X + k\delta Y + \sum_{m,n} C_{mn}^{(k)} X^m Y^n$$
$$Y_k = Y + \sum_{m,n} D_{mn}^{(k)} X^m Y^n \tag{3.24}$$

and the formal differential system

$$\frac{dX}{dk} = \delta Y + \sum_{m,n} a_{mn} X^m Y^n \quad ; \quad \frac{dY}{dk} = \sum_{m,n} b_{mn} X^m Y^n \tag{3.25}$$

The mapping (3.23) can be associated with a differential equation of the type

$$\frac{d^2x}{dt^2} = f(x, \frac{dx}{dt}) \qquad (3.26)$$

In the next chapter we will present some applications of the method : the study of non linear parametric resonance ; the study of periodic hamiltonian systems.

APPENDIX

Details of calculations for the determination of the point mapping associated with a differential equation of the form :

$$\frac{d^2x}{dt^2} + a^2x = f(x,\frac{dx}{dt})$$

Let us start with the following complex mapping τ

$$x_1 = sx + \sum_{m,n} C_{mn}x^m y^n$$

$$y_1 = s^{-1}y + \sum_{m,n} C^*_{mn}x^n y^m \qquad s = \exp[jw] \qquad (3.27)$$

The series x_k, y_k (coordinates of the kth consequent by the point mapping of (x,y)) are determined by writting $\tau^{k+1}=\tau.\tau^k$.This leads to a system of difference equations for the C_{mn} coefficients. For the second order terms :

$$C_{20}^{(k+1)} - sC_{20}^{(k)} = s^{2k}C_{20}$$

$$C_{11}^{(k+1)} - sC_{11}^{(k)} = C_{11} \qquad (3.28)$$

$$C_{02}^{(k+1)} - sC_{02}^{(k)} = s^{-2k}C_{02}$$

For the third order terms

$$C_{30}^{(k+1)} - sC_{30}^{(k)} = 2C_{20}s^kC_{20}^{(k)} + C_{11}s^kC_{02}^{*(k)} + s^{3k}C_{30}$$

$$C_{21}^{(k+1)} - sC_{21}^{(k)} = 2C_{20}s^kC_{11}^{(k)} + C_{11}s^kC_{11}^{*(k)} + C_{11}s^{-k}C_{20}^{(k)} + 2C_{02}s^{-k}C_{02}^{*(k)} + s^kC_{21}$$

$$C_{12}^{(k+1)} - sC_{12}^{(k)} = 2C_{20}s^kC_{02}^{(k)} + C_{11}s^kC_{20}^{*(k)} + C_{11}s^{-k}C_{11}^{(k)} + 2C_{02}s^{-k}C_{11}^{*(k)} + s^{-k}C_{12}$$

$$C_{03}^{(k+1)} - sC_{03}^{(k)} = C_{11}s^{-k}C_{02}^{(k)} + 2C_{02}s^{-k}C_{20}^{*(k)} + s^{-3k}C_{03} \qquad (3.29)$$

These equations are of the general form

$$C^{(k+1)} - sC^{(k)} = ck^n s^{mk} \qquad n,m \text{ integers} \qquad (3.30)$$

By making the change of variable $C^{(k)} = y^{(k)}s^{mk}$, in (3.30), one gets in terms of the y variable

$$y^{(k+1)} - s^{1-m}y^{(k)} = ck^n s^{-m}$$

Two cases are to be considered :

1) $m\neq 1$. The solution $y^{(k)}$ is in the form

$$y^{(k)} = y_0k^n + y_1k^{n-1} + \ldots y_n$$

The coefficients y_i being fixed by identification of equal powers of k in (3.31). The solution of (3.30) is then

$$C^{(k)} = P_n(k)s^{mk} + As^k$$

where $P_n(k)$ is a n degree polynomial in k and As^k the general solution of equation without second member.

2) m=1. In this case, the solution of (3.31) is a (n+1) degree polynomial in k_1 fixed as before. The solution of (3.30) is

$$C^{(k)} = P_{n+1}(k)s^k + Bs^k$$

The equations used for the determination of the second and third degree coefficients are given below with their solutions

$$C^{(k+1)} = sC^{(k)} \quad \left| \begin{array}{l} C_1 s^k \ — \ C^{(k)} = C_1 k s^{k-1} + a_1 s^k \\ C_2 \ — \ C^{(k)} = C_2(1-s)^{-1} + a_2 s^k \\ C_3 s^{mk}, m \neq 1 \ — \ C^{(k)} = C_3 s^{mk}(s^m - s)^{-1} + a_3 s^k \end{array} \right.$$

a_1, a_2, a_3 are arbitrary constants, fixed by the initial conditions, which are obviously for (3.28) and (3.29) : $C_{mn}^{(1)} = C_{mn}$

A.1. Expressions of the $C_{mn}^{(k)}$ coefficients

2nd order terms :

$$C_{20}^{(k)} = C_{20}\frac{(s^{2k} - s^k)}{s^2 - s} \ ; \ C_{11}^{(k)} = C_{11}\frac{1-s^k}{1-s} \ ; \ C_{02}^{(k)} = C_{02}\frac{s^{-2k} - s^k}{s^{-2} - s} \qquad (3.32)$$

3rd order terms :

$$C_{30}^{(k)} = (C_{30} + A_2 + A_0)\frac{s^{3k} - s^k}{s^3 - s} - A_2\frac{s^{2k} - s^k}{s^2 - s} - A_0\frac{1-s^k}{1-s}$$

$$A_2 = 2C_{20}^{\ 2}(s^2 - s)^{-1} \ , \ A_0 = C_{11}\overset{*}{C}_{02}(s^{-2} - s^{-1})^{-1} \qquad (3.33)$$

$$C_{21}^{(k)} = (C_{21} + B_2 + B_0 + B_{-2})ks^{k-1} - B_2\frac{s^{2k} - s^k}{s^2 - s} - B_0\frac{1-s^k}{1-s} - B_{-2}\frac{s^{-2k} - s^k}{s^{-2} - s}$$

$$B_2 = 2C_{20}C_{11}(1-s)^{-1}, \ B_0 = C_{11}\overset{*}{C}_{11}(1-s^{-1})^{-1} + C_{11}C_{20}(s^2 - s)^{-1}, B_{-2} = 2C_{02}\overset{*}{C}_{02}(s^2 - s^{-1})^{-1}$$

$$C_{12}^{(k)} = (C_{12} + C_2 + C_0 + C_{-2})\frac{s^{-k} - s^k}{s^{-1} - s} - C_2\frac{s^{2k} - s^k}{s^2 - s} \ C_0\frac{1-s^k}{1-s} - C_{-2}\frac{s^{-2k} - s^k}{s^{-2} - s}$$

$$C_2=2C_{20}C_{02}(s^{-2}-s)^{-1},\quad C_0=C_{11}C^{*}_{20}(s^{-2}-s^{-1})^{-1}+C^{2}_{11}(1-s)^{-1},C_{-2}=2C_{02}C^{*}_{11}(1-s^{-1})^{-1}$$

$$C_{03}^{(k)}=(C_{03}+D_0+D_{-2})\frac{s^{-3k}-s^{k}}{s^{-3}-s}-D_0\frac{1-s^{k}}{1-s}-D_{-2}\frac{s^{-2k}-s^{k}}{s^{-2}-s}$$

$$D_0=C_{11}C_{02}(s^{-2}-s)^{-1},D_{-2}=2C_{02}C^{*}_{20}(s^{-2}-s^{-1})^{-1}$$

A.2. Determination of the C_{mn} coefficients

Recall the differential equation

$$\frac{dX}{dk}=-ja_ih_iX+ja_ih_i\sum_{m,n}b_{mn}X^mY^n$$

$$m+x\geq 2$$

$$\frac{dY}{dk}=ja_ih_iY-ja_ih_i\sum_{m,n}b^{*}_{mn}X^mY^n$$

to which is associated the mapping (3.27). First we have $s=\exp[-ja_ih_i]$ and for the second degree terms

$$C_{20}=b_{20}(s-s^2),\quad C_{11}=b_{11}(1-s),\quad C_{02}=b_{03}(s^{-2}-s)/3$$

the third degree terms

$$C_{30}=\frac{s-s^3}{2}\left[b_{30}-\frac{A_2}{s^2-s}+\frac{A_0}{1-s}\right]-A_2-A_0$$

$$C_{21}=ja_ih_is\left[b_{21}-\frac{B_2}{s^2-s}+\frac{B_0}{1-s}+\frac{3B_{-2}}{s^{-2}-s}\right]-B_2-B_0-B_{-2}$$

$$C_{12}=\frac{s^{-1}-s}{2}\left[b_{12}-\frac{C_2}{s^2-s}+\frac{C_0}{1-s}+\frac{3C_{-2}}{s^{-2}-s}\right]-C_2-C_0-C_{-2} \qquad (3.34)$$

$$C_{03}=\frac{s^{-3}-s}{4}\left[b_{03}+\frac{D_0}{1-s}+\frac{3D_{-2}}{s^{-2}-s}\right]-D_0-D_{-2}$$

where the A_i,B_i,C_i,D_i are functions of the 2nd order coefficients.

CHAPTER IV
PARAMETRIC RESONANCE. CONSERVATIVE SYSTEMS

In this chapter some applications of the method described in chapter III are given.
In the first section, we deal with the phenomena of parametric resonance which may occur in the field of non linear differential equations depending on periodic coefficients for instance. Such equations are encountered in such varied domains as electronics (parametric amplifiers) physics (behaviour of particles in an alternating gradient accelerator (Ref. 43)) and astronautics (oscillations of a satellite around its center of mass (Ref. 44)). More generally, studying the stability of periodic solutions of any differential system (autonomous or not) by the perturbation method involves differential equations with periodical coefficients.

The second section of the chapter deals with conservative systems or more precisely, periodic hamiltonian systems. A numerical method of studying these systems is given. This involves defining an approximating mapping.

IV.1. Parametric Resonance

For the sake of simplicity, let us begin with the following differential equation :

$$\frac{d^2x}{dt^2} + a^2x = f(x, \frac{dx}{dt}, t) \qquad (4.1)$$

where $f(0,0,t)=0$ (the origin is an equilibrium point), f being periodic with respect to t (period equal to T) and satisfying the classical conditions on existence and uniqueness of solutions. It is well known that, when the natural period $2\pi/a$ and the period of excitation T are in rational ratio, what is called parametric resonance may occur. It is the aim of this section to study this phenomenon, i.e., the influence of resonance upon the nature and the stability of the equilibrium point, by means of the associated mapping.

IV.1.1. The Exceptional Cases - $w = 2k\pi/q$

We saw in the previous chapter that the mapping associated to equation (4.1) is written, in terms of complex variables :

$$\begin{aligned} X_1 &= \exp[jw]X + \sum_{m,n} C_{mn}X^mY^n \\ & \qquad\qquad\qquad m+n \geq 2 \qquad (4.2) \\ Y_1 &= \exp[-jw]Y + \sum_{m,n} C^*_{mn}X^nY^m \end{aligned}$$

with $w = aT$. Furthermore we saw in the chapter on the critical case with complex multipliers that when w is commensurable with 2π, the classical method using a Lyapunov function V can no longer be applied since the fact that some of the coefficients appearing in the expression of ΔV are null makes it im-

possible to reduce the variation ΔV to a sign definite form. These cases were called exceptional cases.
In(Ref.45,47)the following method is used:a first step consists in a non linear change of variables (non singular) which is performed on (4.2) in order to eliminate the maximum of terms, thus showing up the terms which are essential for the study. In a second step the nature and the stability of the singular point (origin) is determined by means of a study in terms of polar coordinates. Here a slightly different form is given which directly relates the exceptional case to the critical case with two multipliers equal to one (linear part diagonal). Indeed, for $w = 2k\pi/q$, taking the qth power of the mapping (4.2), one gets :

$$X_q = X + \sum_{m,n} C_{mn}^{(q)} X^m Y^n$$
$$Y_q = Y + \sum_{m,n} C_{mn}^{*(q)} X^n Y^m \tag{4.3}$$

It is clear, since none of the terms of (4.3) can be cancelled by performing a change of variable, that (4.3) contains the terms relevant for the study. The other terms have been cancelled by taking the qth power of the mapping. This avoids performing the non linear change of variables mentioned above. Let us now consider the most frequent cases that may occur, that is the cases in which q is relatively small.

IV.1.2. $q = 2$
Considering the expressions (3.32) it is easy to see that

$$C_{20}^{(2)} = C_{11}^{(2)} = C_{02}^{(2)} = 0$$

that is the second degree terms become null in the expression of the second power of the mapping, while the third order terms remain and thus

$$X_2 = X + \sum_{m+n=3}^{\infty} C_{mn}^{(2)} X^m Y^n$$
$$Y_2 = Y + \sum_{m+n=3}^{\infty} C_{mn}^{*(2)} X^n Y^m \tag{4.4}$$

This section is continued as in section II.4.2. Here because of the number of remaining parameters (8, the coefficients of the third degree terms) nothing general can be said. However in the following two exceptional cases more specific results can be obtained.

IV.1.3. $q = 3$
From (3.32) we get

$$C_{20}^{(3)} = C_{11}^{(3)} = 0$$
$$C_{02}^{(3)} = 3\exp[2jw]C_{02} \quad (\neq 0 \text{ if } C_{02} \neq 0) \tag{4.5}$$

For $C_{02} \neq 0$

$$X_3 = X + 3\exp[2jw]C_{02}Y^2 + \sum_{m+n=3}^{\infty} C_{mn}^{(3)}X^mY^n$$

$$Y_3 = Y + 3\exp[-2jw]C_{02}^{*}X^2 + \sum_{m+n=3}^{\infty} C_{mn}^{*(3)}X^nY^m \quad (4.6)$$

written in terms of complex variables. If the following polar change of variable $X = r\exp[jv]$, $Y = r\exp[-jv]$ is performed on (4.6) then it becomes, with the notation of section (II.4.2.)

$$r_3 = r + P_2(v)r^2 + \dots$$

$$v_3 = v + Q_2(v)r + \dots$$

where :

$$P_2(v) = R_e[3C_{02}\exp(-3jv+2jw)] = 3\cos[3v-2w-\tau_{02}]|C_{02}|$$

$$Q_2(v) = I_m[3C_{02}\exp(-3jv+2jw)] = 3\sin[3v-2w-\tau_{02}]|C_{02}|$$

where $|.|$ means absolute value of . ; $R_e[.]$ real part of [.], $I_m[.]$ imaginary part of [.] and τ_{02} is the argument of C_{02}. The invariant directions are given by :

$$\sin[3v-2w-\tau_{02}] = 0 \;; \qquad \text{i.e.,} \qquad v = \frac{2w+\tau_{02}+k\pi}{3}$$

It is easy to see that these six directions are alternatively stable and unstable, repulsive and attractive for the discrete trajectories thus delimiting six hyperbolic sectors (Fig. 4.1). The fixed point origin is therefore an unstable hyperbolic point

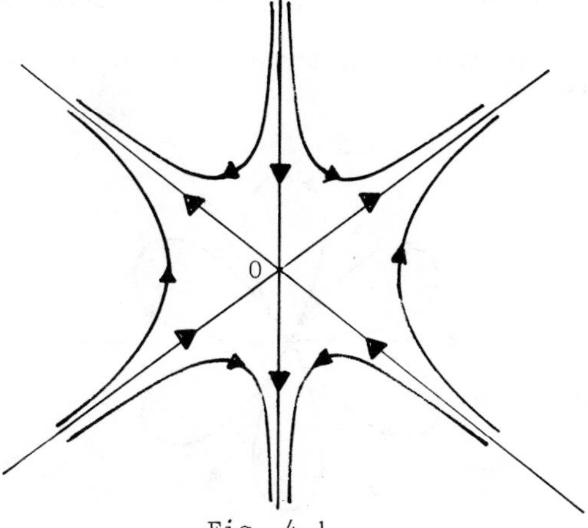

Fig. 4.1

Let us now consider the case when $C_{02} = 0$. From (3.33) we have

$$C_{30}^{(3)} = C_{12}^{(3)} = C_{03}^{(3)} = 0$$

$C_{21}^{(3)} \neq 0 \qquad (\text{if } C_{21} \neq 0)$

Then (4.6) becomes

$$X_3 = X + C_{21}^{(3)} X^2 Y + \ldots \; ; \quad Y_3 = Y + C_{21}^{*(3)} XY^2 + \ldots \tag{4.7}$$

and, in polar coordinates :

$$r_3 = r + R_e[C_{21}^{(3)}]r^3, \ldots \; ; \; v_3 = v + I_m[C_{21}^{(3)}]r^2 + \ldots \tag{4.8}$$

with the obvious result that the origin is stable if $R_e[C_{21}^{(3)}] < 0$. If $I_m[C_{21}^{(3)}] \neq 0$ the fixed point is a focus. If $I_m[C_{21}^{(3)}] = 0$ there may be invariant directions depending on higher degree terms. If $C_{21} = 0$ the same method is used. As an example let us consider the mapping :

$$\begin{aligned} x_1 &= x \cos w - y \sin w + x^4 + y^4 \\ y_1 &= x \sin w + y \cos w \end{aligned} \qquad \text{with } w = \frac{2\pi}{3} \tag{4.9}$$

that, written in complex variables , gives

$$X_1 = \exp[jw]X + \frac{1}{8}(X^4 + 6X^2Y^2 + Y^4) \; ; \; Y_1 = \exp[-jw]Y + \frac{1}{8}(X^4 + 6X^2Y^2 + Y^4) \tag{4.9'}$$

Taking the third power of the mapping, one gets

$$X_3 = X + 3\exp[2jw]X^4/8 \qquad ; \quad Y_3 = Y + 3\exp[-2jw]Y^4/8 \tag{4.10}$$

that is, in polar coordinates

$$r_3 = r + 3/8 \cos(3v+2w)r^4 + \ldots \qquad ; \quad v_3 = v + 3/8 \sin(3v+2w)r^3 + \ldots \tag{4.10'}$$

We get six invariant directions, alternately stable and unstable, delimiting six elliptic sectors

Fig. 4.2.

IV.1.4. q = 4

Similarly, in the expressions of the fourth power of the mapping it can be seen that

$$C_{20}^{(4)} = C_{11}^{(4)} = C_{02}^{(4)} = C_{30}^{(4)} = C_{12}^{(4)} = 0$$

and

$$C_{21}^{(4)} \neq 0 \ , \quad C_{04}^{(4)} \neq 0$$

in general if second or third degree terms appear in the expressions of the mapping (4.2) .

Thus

$$\begin{aligned} X_4 &= X + C_{21}^{(4)} X^2 Y + C_{03}^{(4)} Y^3 + \dots \\ Y_4 &= Y + C_{21}^{*(4)} XY^2 + C_{03}^{*(4)} X^3 + \dots \end{aligned} \tag{4.11}$$

and

$$\begin{aligned} r_4 &= r + R_e[C_{21}^{(4)}]r^3 + R_e[C_{03}^{(4)} \exp(-4jv)]r^3 + \dots \\ &= r + R_e[c_{21}^{(4)}]r^3 + |C_{03}^{(4)}| \cos(4v-\tau_{03})r^3 + \dots \\ v_4 &= v + I_m[C_{21}^{(4)}]r^2 + I_m[C_{03}^{(4)} \exp(-4jv)]r^2 + \dots \\ &= v + I_m[C_{21}^{(4)}]r^2 - |C_{03}^{(4)}| \sin(4v-\tau_{03})r^2 + \dots \end{aligned} \tag{4.12}$$

with τ_{03} = argument of $C_{03}^{(4)}$

The invariant directions are given by solving the equation :

$$Q_3(v) = I_m[C_{21}^{(4)}] - |C_{03}^{(4)}| \sin(4v-\tau_{03}) = 0 \tag{4.13}$$

Obviously, such an equation has roots, if and only if

$$a = \frac{I_m[C_{21}^{(4)}]}{|C_{03}^{(4)}|} \leqq 1$$

Different cases are considered :

1°) a>1. No invariant directions exist and the stability of the fixed point is determined by the sign of

$$h_0 Q_3(v) = Q_3(v)\int_0^{2\pi} \frac{P_3(v)}{Q_3(v)} dv, \text{ with } P_3(v) = R_e[C_{21}^{(4)}] + |C_{03}^{(4)}| \cos(4v-\tau_{03}).$$

It can be easily shown that the sign of $h_0 Q_3(v)$ is the same as that of $R_e[C_{21}^{(4)}]$ thus

$$R_e[C_{21}^{(4)}] < 0 \ (>0) \Longrightarrow 0 \text{ stable (unstable) focus.}$$

When $C_{21}^{(4)} = 0$, higher degree terms must be taken.

2°) $a \leqq 1$. There are eight invariant directions, superimposed in pairs for certain values of a such as $a=1, \sqrt{2}/2$ (this corresponds to a root of multiplicity 2 for equation (4.13)) and the stability is fixed by the sign of $P_3(v)$ on these directions.
Putting $a = \sin\alpha$, the invariant directions are given by

$$4v - \tau_{03} = \alpha + 2k\pi \quad \text{(four directions)}$$

$$4v - \tau_{03} = \pi - \alpha + 2k\pi \quad \text{(four directions)}$$

that is, two families of invariant directions each containing four directions equally spaced. For each member of the family the stability is given by the sign of

$$R_e[C_{21}^{(4)}] + |C_{03}^{(4)}|\cos\alpha \qquad \text{for the first one}$$

$$R_e[C_{21}^{(4)}] - |C_{03}^{(4)}|\cos\alpha \qquad \text{for the second one}$$

Figure 4.3 shows the different results when $R_e[C_{24}^{(4)}]<0$ and $a<1$.

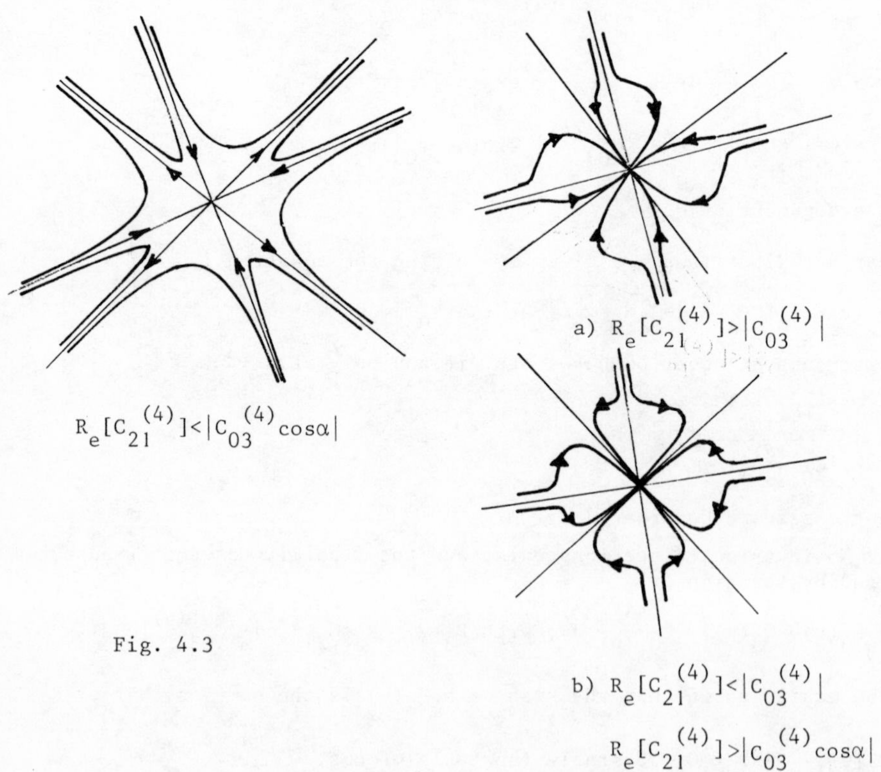

Fig. 4.3

The case when $R_e[C_{21}^{(4)}] = |C_{03}^{(4)}\cos\alpha|$ leading $P_3(v) = 0$ on a family of directions needs to take into account higher degree terms.

IV.1.5. Generalisation (Ref. 33)
Let us now consider the general problem $w=2k\pi/q$, k and q are integers relatively prime and q not necessarily equal to 2, 3 or 4. The qth consequent of the mapping (4.2) is expressed by equations (4.3) and the $C_{mn}^{(q)}$ coefficients are determined by solving linear difference equations of the form :

$$C_{mn}^{(k+1)} - \exp[jw]C_{mn}^{(k)} = \exp[jq(m-n)w]C_{mn} + P_{mn} , \quad C_{mn}^{(1)} = C_{mn} \tag{4.14}$$

where (see section III.1.2) P_{mn} is constituted by terms coming from

$$C_{ij}x_k^{\ i}y_k^{\ j} , \quad i,j \text{ integers and } i+j<m+n$$

i.e., P_{mn} depends on coefficients $C_{ij}^{(q)}$, $C_{ij}^{*(q)}$ of terms of lower degree than $N=m+n$ and thus, having previously been determined by means of equations like (4.14) corresponding to the terms of lower degree than N. Furthermore it is shown in (Ref. 33) that the degree of P_{mn} with respect to the variables $\exp(jw)$, $\exp(-jw)$ and k is at most equal to N. The solution of (4.14) is

$$C_{mn}^{(q)} = C_{mn}\frac{\exp(jqw)-\exp(jq(m-n)w)}{\exp(jw)-\exp(j(m-n)w)} + Q_{mn} \quad \text{if } m\neq n+1 \tag{4.15}$$

$$C_{mn}^{(q)} = qC_{mn}\exp(j(q-1)w) + Q'_{mn} \quad \text{if } m=n+1 \tag{4.16}$$

where Q_{mn}, Q'_{mn} are polynomials in $\exp(jw)$, $\exp(-jw)$, k of degree at most equal to N.

In the case when $w=2k\pi/q$, all the terms $[\exp(jqw)-\exp(j(m-n)qw)][\exp(jw)-\exp(j(m-n)w)]^{-1}$ are zero, except for the terms such as

$$m-n-1 = k'q, \text{ whatever } k' \text{ integer} \tag{4.17}$$

These terms are therefore expressed as : $q\exp[j(q-1)w]$. For $k'=0$ in equation (4.17) one gets the terms (4.16). It should be noted that the lowest degree terms (other than those with $k'=0$) which satisfy (4.17) are $m=0$, $n=q-1$. This enables us to write

$$\begin{aligned} X_q &= X + \sum_{i=1}^{\ell} C_{i+1,i}^{(q)}X^{i+1}Y^{i} + C_{0,q-1}^{(q)}Y^{q-1} + \dots \\ Y_q &= Y + \sum_{i=1}^{\ell} C_{i+1,i}^{*(q)}X^{i}Y^{i+1} + C_{0,q-1}^{*(q)}X^{q-1} + \dots \end{aligned} \tag{4.18}$$

where ℓ is an integer such that $2\ell+1 \leqq q-1$; the dotted lines represent terms of higher degree than $(q-1)$. In equation (4.18) two types of terms can be distinguished :

- the terms $X^{i+1}\ Y^{i}$: these can be called non resonant terms since they cannot be eliminated, even when $w\neq 2k\pi/q$ by a non linear change of variables
- the terms Y^{q-1} : the resonant terms.

It can easily be seen from the form of equations (4.18) that the high degree resonance (q high) becomes more and more unlikely, when increasing q because of the non resonant terms. All that can be said is that, except for very special cases, a q order resonance may appear (giving rise to instability) if the mapping does not have non linear terms of lower degree than

q-2 if q is odd

q-1 if q is even

IV.1.6. Example

Let us consider the following differential equation which will be taken up again in the next section :

$$\frac{d^2x}{dt^2} + x + f(t)(b_1x^2+b_2x^3) = 0 \tag{4.19}$$

where f(t) is the signal :

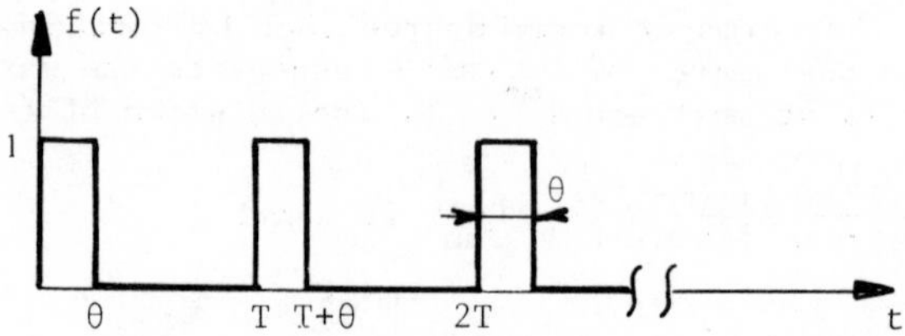

Using the results obtained in chapter III to calculate the coefficients of the mapping associated with (4.19) one gets the following expressions written in terms of $X=x+j\frac{dx}{dt}$, $Y=x-j\frac{dx}{dt}$

$$X_1 = \exp(-jT)X + \sum_{mn} C_{mn}X^mY^n, \quad X=X(nT), \quad Y=Y(nT)$$

$$Y_1 = \exp(jT)Y + \sum_{mn} C^*_{mn}X^nY^m, \quad X=X[(n+1)T], \quad Y=Y[(n+1)T] \tag{4.20}$$

with

$$C_{20} = \exp(-jT)\,[\exp(-j\theta)-1]\frac{b_1}{4}, \quad C_{11} = \exp(-jT)\,(1-\exp(j\theta))\,\frac{b_1}{2}$$

$$C_{02} = \exp(-jT)\,(1-\exp(3j\theta))\frac{b_1}{12}$$

$$C_{30} = \exp(-jT)\,[\exp(j\theta)(\exp(-j\theta)-1)^3\,\frac{b_1^2}{24} + (\exp(-2j\theta)-1)\,\frac{b_2}{16}\,]$$

$$C_{21} = \exp(-jT)\,[(30j\theta-\exp(3j\theta)-8-9\exp(j\theta)+18\exp(-j\theta))\,\frac{b_1^2}{72} - 3j\theta\,\frac{b_2}{8}\,]$$

$$C_{12} = \exp(-jT)\,[(\exp(+j\theta)-1-3\exp(-j\theta)+5\exp(2j\theta)-2\exp(3j\theta))\,\frac{b_1^2}{24}$$

$$-3(\exp(2j\theta)-1)\frac{b_2}{16}]$$

$$C_{03} = \exp(-jT)\left[(\exp(4j\theta)-2\exp(3j\theta)-1+2\exp(j\theta))\frac{b_1^2}{48} + (\exp(4j\theta)-1)\frac{b_2}{32}\right]$$

For $T=2\pi/3$, provided that $b_1 \neq 0$, the origin is an hyperbolic point with 6 invariant directions as seen above. For $T=2\pi/4$, the conditions for the existence of invariant curves passing through the origin i.e., $|a|<1$ (section IV.1.4), have been determined numerically by computing the fourth power of the mapping and then calculating for various values of b_1 and b_2 the coefficient a. These results give the curves of Fig. 4.4, which is drawn with $wb_1=1$, and $w=2\pi/4$.

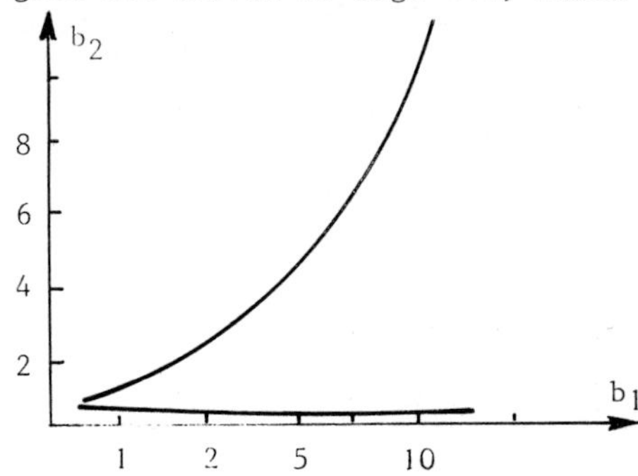

Fig. 4.4

The region $|a|<1$ is the region of instability of the origin. Indeed, the equation (4.19) is conservative or, more exactly, it derives from an Hamiltonian causing the equilibrium points to be either elliptic ($|a|>1$) or hyperbolic points.

Let us now consider for the case $b_1=0$, $w=2\pi/4$, the influence of the function f(t) on the appearance of resonance in order to test the following conjecture:

"For a differential equation of the form $\frac{d^2x}{dt^2} + x = f(t)x^3$ where f(t) is a periodic function of period equal to $\pi/2$, resonance (instability of the equilibrium point origin) occurs if and only if

$$|h| < A/6$$

where h is the mean value of f(t) and A the amplitude of the first harmonic"

The conjecture has been numerically tested for $f(t)= h+A \sin 4t$, making an approximation of f(t) by a piecewise constant function and determining numerically the coefficients of the third order terms of the associated mapping. As regards the value of a, it would appear, roughly speaking, that

$$a = \frac{6h}{A} \tag{4.21}$$

For piecewise constant functions f(t) (with not too many constant subintervals) it is easy to determine the coefficients in closed form. For instance for the above example, one gets

$$C_{21} = -3j\exp(-jT)\,\theta\frac{b_2}{8} \qquad \text{and } C_{24}^{(4)} = -12j\theta\,\frac{b_2}{8}$$

$$C_{03} = \exp(-jT)\,(\exp(4j\theta)-1)b_2/32 \text{ and } C_{03}^{(4)} = (\exp(4j\theta)-1)\,\frac{b_2}{8}$$

thus

$$a = \frac{I_m |C_{21}^{(4)}|}{|C_{03}^{(4)}|} = \frac{12\theta}{\sqrt{2(1-\cos 4\theta)}}$$

If A_1, B_1 represent the coefficients of the first harmonic terms of the Fourier series expansions of $f(t)[=h+A_1\cos 4t+B_1\sin 4t]$ we get

$$h = 2\theta/\pi, \quad A_1 = \sin 4\theta/\pi, \quad B_1 = (1-\cos 4\theta)/\pi$$

and thus

$$a = 6h(A_1^2 + B_1^2)^{-1/2}$$

Let us now consider the following case :

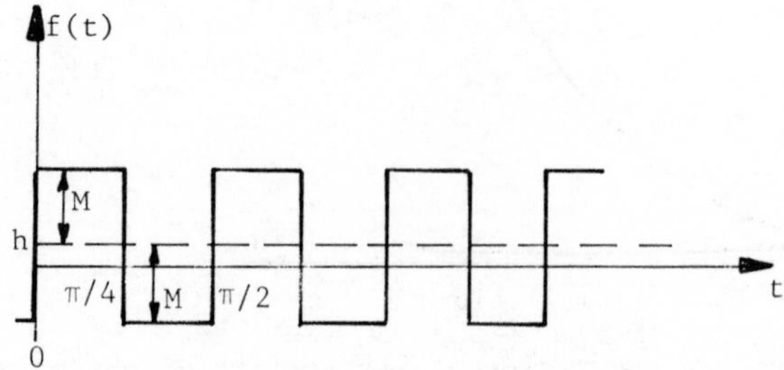

The coefficients C_{21} and C_{03} are easily computed, one gets $C_{21} = \frac{2\pi b_2 h}{16}$, $C_{03} = j\frac{b_2 M}{8}$ and $a = \frac{3\pi}{2}\frac{h}{M}$. Here $A_1=0$, $B_1 = \frac{4M}{\pi}$ which satisfy equation (4.21).

For the following function :

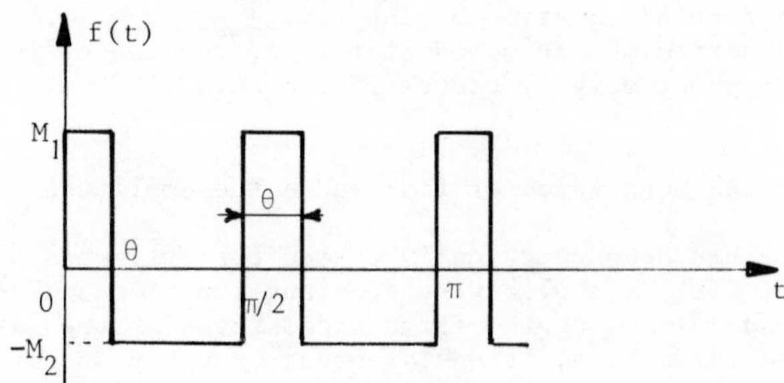

one gets $C_{21} = \frac{3jb_2}{8}(\theta M_1 - (\frac{\pi}{2} - \theta)M_2)$, $C_{03} = j\frac{b_2}{32}(M_1 - M_2)[(1-\cos 4\theta) + j\sin 4\theta]$ and

$a = 6\sqrt{2}\,\frac{\theta M_1 + (\frac{\pi}{2} - \theta)M_2}{(M_1 - M_2)\sqrt{1-\cos 4\theta}}$. Finally $h = \frac{2}{\pi}[\theta M_1 - (\frac{\pi}{2} - \theta)M_2]$, $A_1 = \frac{1}{\pi}(M_1 - M_2)\sin 4\theta$,

$B_1 = \frac{1}{\pi}(M_1 - M_2)(1-\cos 4\theta)$ which satisfy equation (4.21).

IV.2. Conservative Systems

This section is devoted to second order non linear conservative point mappings

$$x_1 = f(x,y) \quad , \quad y_1 = g(x,y) \tag{4.22}$$

where conservative means that (4.22) has the invariant integral :

$$\iint_D I(x,y)dx\,dy \tag{4.23}$$

We will restrict ourselves to the case of area preserving mappings. That is those for which $\iint_D dxdy$ is invariant by (4.22). Obviously a sufficient and necessary condition for (4.22) to be area preserving is that : $\det[J]=1$, J being the Jacobian matrix of the mapping :

$$J = \begin{bmatrix} \frac{\partial f}{\partial x} & \frac{\partial f}{\partial y} \\ \frac{\partial g}{\partial x} & \frac{\partial g}{\partial y} \end{bmatrix}$$

First, it should be noted that the restriction on (4.22), to be area preserving, is not very restrictive since, given the analyticity of I(x,y) it is possible with a non linear change of variables : $u=u(x,y)$, $v=v(x,y)$ to make the mapping (4.22), expressed in terms of the u,v variables, area preserving. Before starting on the main part of this section, it is worthwhile to give a brief survey of some of the problems which arise in conservative dynamical systems in order to emphasize the importance of developing numerical methods for studying them.

IV.2.1. Area Preserving Point Mappings. Specific Problems

Let us consider the following area preserving point mapping T

$$\begin{aligned} x_1 &= y + f(x) \\ y_1 &= -x + f(x_1) \end{aligned} \tag{4.24}$$

where $f(x)= ax+(1-a)x^2$. This mapping (or very similar ones) has been extensively studied in the literature (Ref.42,48-50). There are two reasons for this:

1 - Its simplicity : it is a quadratic mapping (simpler form of non linearity) depending on one parameter in such a way its fixed points (double points : O:x=0, y=0 ; A:x=1, y=0) are independent of the parameter.

2 - Its generality : in spite of its simplicity, it retains the properties of the general cases when a is varied.

For a=0.125, Fig. 4.5 is obtained. It can be done the following qualitative description. The origin is a center (elliptic point) surrounded by apparently closed invariant curves and island structures. The figure shows a nine order island structure : nine elliptic points, fixed points of the ninth power of the mapping separated by nine saddle points, also fixed points of T^9. Their invariant curves define a cell around each elliptic point. If greater precision could be achieved it would be possible to see that each cell (each island) exhibits the same features as the domain around the origin that is closed curves, island structures... and so on.

The point A(x=1,y=0) is a saddle. If we consider the two leftmost invariant curves passing through this point it can be seen that they do not join smoothly at point B but cut across each other with a non null angle. This causes them to begin to oscillate, resulting in an infinity of points of intersection. These points are called homoclinic points. The curves are then caught in an "endless dance" (see (51), for a very exciting description of this phenomenon!)

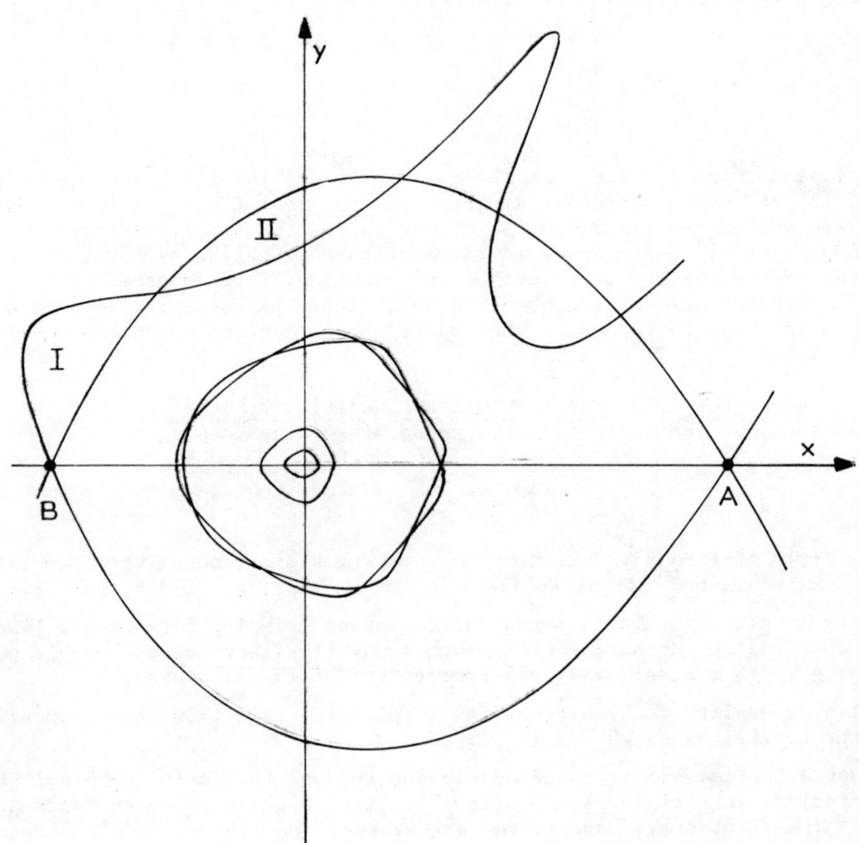

Fig. 4.5

Let (I) represent the outside loops and (II) the inside loops (Fig. 4.5). The area preserving property means that all inside loops and all outside loops have equal area and, in addition, for reasons of symmetry of the mapping the outside loop area is equal to inside loop area. However this last fact is not crucial. Mac Millan (51) gives a vivid description of the behaviour of the loops:

> "Outside loops become long, thin tentacles reaching toward infinity inside loops become "worms" curled up in the interior. The development of this pattern, and the ingenuity displayed by the "worms" in avoiding contacts with others of the same family are fascinating to watch".

If the invariant curves passing through point B had joined smoothly at the first meeting point B, they would have made a single closed loop around the stable fixed point origin, and all iterates of any point inside the loop would remain inside. In fact, the area point inside the loop is reduced by the inside loops, that seem to approach the stable fixed point though more and more slowly.

The (x,y) plane can be roughly divided in two parts. In a domain D around the fixed point origin, the behaviour of the solutions (iterates of 4.24) is quite regular, i.e., the iterates seems to remain (at least for a very long time) on closed curves, the system being apparently integrable. Outside D, further from the stable fixed point origin, the behaviour becomes irregular, it is said to be stochastic. Birkhoff has shown that in the vicinity of every homoclinic points, there is an infinity of high order cycles, all unstables : this causes the stochastic behaviour. In fact, for the example chosen here, there is a region of stochastic instability, that is, the iterates finally escape to infinity. But this is not always the case and it is possible to come accross examples with a bounded zone of stochasticity.

One of the main problem is that of finding the limit of the domain of stability, that is finding the domain of integrability (or pseudo-integrability).

This question could be answered by finding (and proving its existence if possible) the closed invariant curve which is furthest from the origin since this would set the boundary of the stability domain. If such an invariant curve would be shown to exist it would follow that the iterates of all points inside would remain inside.

The K.A.M. (Kolmogorov - Arnold - Moser) theory has stated that, provided the rotation number at the origin (the argument of the multiplier) is sufficiently far from any rational number, then there are analytic closed invariant curves in a sufficiently small neighbourhood of the origin. From a practical point of view, however, this result fails to give a sufficiently realistic estimate of the domain of stability. Up until now, therefore, numerical studies are the only ones to have, even partially, answered this question.

IV.2.2. The Numerical Studies

It is clear that the point mapping transformation is well suited to numerical use. Generally, however, the physical problems cannot be formulated directly by means of point mapping. The study of conservative dynamical systems with two degrees of freedom (Hamiltonian formulation in the case of the three-body problem in astronomy, for instance) can be reduced, by the means of a surface of section (first introduced by Poincaré) to the study of a mapping of the surface onto itself. The fourth order, autonomous differential system can be reduced, using the integral of motion constituted by the Hamiltonian to a third order differential system and then to a second order point mapping by considering the two consecutive intersections of a space trajectory with the surface of section. Generally such a reduction is a purely formal one in the sense that one cannot avoid numerical integration. However, this idea is still useful for representing the properties of the solutions : for instance, a periodic solution is represented by a fixed point, or a cycle of the mapping. Of course, it would be much more fruitful if the analytical expressions of the mapping could be obtained since it would result in time saving and would reduce the amount of rounding off errors at each step of the numerical integration.

It is the aim of the following section to provide a partial solution to this problem by considering the case of periodic Hamiltonian systems.

IV.2.3. Periodic Hamiltonian Systems

We consider here the systems that can be formalised by the equations :

$$\frac{dp}{dt} = \frac{\partial H(p,q,t)}{\partial q}$$
$$\frac{dq}{dt} = -\frac{\partial H(p,q,t)}{\partial p} \qquad (4.25)$$

where H is periodic in t with period T. It has been known since Poincaré that a point mapping can be associated with system (4.25), relating the values of p,q at time t=nT to those at t=(n+1)T, p_1,q_1. Thus, we obtain a mapping M of the form

$$M : p_1 = P(p,q) \quad , \quad q_1 = Q(p,q) \qquad (4.26)$$

Provided that H is an analytic function, M is one to one and analytic in the neighbourhood of the origin p=q=0 (we assume that the origin is an equilibrium point for (4.25)). Furthermore, M is area preserving.

Considering a conservative autonomous dynamical system with two degrees of freedom (which, as we said before can be reduced to the study of a 2nd order point mapping) and a periodic trajectory, the motion in a neighbourhood of the periodic trajectory can be described by equation of the type (4.25) where p=q=0 corresponds to the periodic motion in the space (p,q,t) and H is periodic in t with period 2π.

From a practical point of view, however, the analytic expressions of P and Q are generally unknown. There are only some special cases where such expressions can be found. For instance, when the periodic dependence of the hamiltonian is made through ideal Dirac impulses, it is possible to find the mapping. This is the case for (Ref 42)

$$\frac{dx}{dt} = y \;, \quad \frac{dy}{dt} = -x + \sum_{n=-\infty}^{+\infty} \delta(t-nT)x^2 \quad \text{for which } H = \frac{x^2+y^2}{2} - \frac{\Sigma\delta(t-nT)x^3}{3}$$

and the corresponding mapping :

$$x_1 = x\cos T + (y+x^2)\sin T$$
$$y_1 = x\sin T + (y+x^2)\cos T$$

(x,y) being the state just before an impulse, (x_1,y_1) the state just before the next one.

The following example (Ref. 41) is related to the study of the transversal motion of particles in an alternating gradient accelerator :

$\frac{dx}{dt} = y$, $\frac{dy}{dt} = -x + f(t)x^2$ where $f(t)$ is a periodic piecewise constant function. The solution can be obtained by means of elliptic functions on each constant f subinterval and depends on initial conditions. The total transformation is the product of the partial ones, the results being generally very involved.

Chapter III gave a method which enabled some initial terms of the series expansions of P and Q to be found. In general, for a strictly local study, the complete expressions of P and Q are not needed, but only the initial terms as was seen in the last section dealing with parametric resonance. Although taking higher degree terms into account should improve the knowledge of the behaviour of the solution in the vicinity of the origin, the truncation of the series of P and Q could destroy its conservative nature and causes a change in the qualitative properties of the discrete trajectories in the plane p,q.

A method of calculating an approximating mapping for equations of type (4.25) starting from the truncated mapping which is made up of the initial terms of the series expansions of the true mapping M is given below. The approximating mapping is constructed in such a way as to satisfy the following requirements:

1 - have the same initial terms as M. This first requirement is designed to insure the same local behaviour for the solutions of M and the approximating mapping

2 - be area preserving. This is done by adding higher degree terms to the truncated mapping. This second requirement is made in order to increase the size of the domain in which the behaviour of M and the approximating mapping is, at least, qualitatively the same (Ref. 52)

IV.2.4. Preliminary Remarks

This section consists of a brief survey of some results given by Engels (53) dealing with polynomial transformations with a constant Jacobian. These are also called entire Cremona transformations and will be useful for determining the approximating mapping.

a) The following polynomial mapping :

$$\begin{aligned} u_1 &= u + \beta \sum_{n=2}^{N} a_n(\alpha u + \beta v)^n \\ v_1 &= u - \alpha \sum_{n=2}^{N} a_n(\alpha u + \beta v)^n \end{aligned} \tag{4.27}$$

where N is any positive integer higher than 2 ; α,β,a_i are constants, is an area preserving mapping (jacobian unity).

b) If the mapping

$$u_1 = g(u,v) \quad , \quad v_1 = h(u,v) \tag{4.28}$$

is area preserving, then all the mappings of the form

$$\begin{aligned} u_1 &= g(u,v) + \sum_{n=1}^{N} \alpha_n(h(u,v))^n \\ v_1 &= h(u,v) \end{aligned} \tag{4.29}$$

or equivalently :

$$u_1 = g(u,v)$$
$$v_1 = h(u,v) + \sum_{n=1}^{N} \alpha_n [g(u,v)]^n \qquad (4.29')$$

where N is any positive integer, are also area preserving.

IV.2.5. Approximating Mapping. Osculation Up To The Second Degree Terms
Let us assume that, using the method given in chapter III, it has been possible to determine the series expansions of P and Q up to the second degree terms. So that the truncated mapping T_C is expressed, in terms of real variables :

$$x_1 = ax + by + \bar{C}_{20}x^2 + \bar{C}_{11}xy + \bar{C}_{02}y^2$$
$$y_1 = cx + dy + \bar{D}_{20}x^2 + \bar{D}_{11}xy + \bar{D}_{02}y^2 \qquad (4.30)$$

In order to avoid excessive calculations in what follows the mapping T_C is regarded as the product of two mappings T_1 and T_2

$$(x,y) \xrightarrow{T_1} (x',y') \xrightarrow{T_2} x_1,y_1$$

The first T_1 is the linear part of (4.30)

$$T_1 : x' = ax + by \ , \quad y' = cx + dy$$

T_1 is clearly area preserving since ad-bc=1. T_2 has the form :

$$x_1 = x' + \sum_{m+n=2} C_{mn} x'^m y'^n$$
$$y_1 = y' + \sum_{m+n=2} D_{mn} x'^m y'^n \qquad (4.31)$$

where the coefficients C_{mn}, D_{mn} are expressed simply in terms of the coefficients $\bar{C}_{mn}, \bar{D}_{mn}$. Since T_1 is already area preserving the problem is reduced to finding a conservative approximating mapping osculating T_2. Since T_2 is derived from the truncation of an area preserving mapping, some relations can be obtained for the C_{mn}, D_{mn} coefficients, by writing for the full mapping J (jacobian)≡1. From which one gets

$$D_{11} + 2C_{20} = 0$$
$$2D_{02} + C_{11} = 0 \qquad (4.32)$$

Thus out of the six coefficients of T_2, four (at the most) are independent. It is therefore necessary to construct a conservative mapping depending on four parameters. This is easily done from the product of the three mappings (4.28), (4.29), (4.29') given in section IV.2.4.

Consider the following mapping

$$x' = x + \beta(\alpha x + \beta y)^2 = g(x,y)$$
$$y' = y - \alpha(\alpha x + \beta y)^2 = h(x,y)$$

$$x_1 = g(x,y) + \delta h^2(x,y) \tag{4.33}$$

$$y_1 = h(x,y) + \gamma x_1^2 = h(x,y) + \gamma[g(x,y) + \delta h^2(x,y)]^2$$

This mapping, the product of three area preserving mappings is obviously area preserving. It depends on four parameters $\alpha,\beta,\delta,\gamma$ which can be chosen such that its second degree terms are the same as (4.29) by solving

$$\beta\alpha^2 = C_{20} \;;\; 2\alpha\beta = C_{11} \;;\; \beta^3+\delta = C_{02} \;;\; -\alpha^3+\gamma = D_{20} \tag{4.34}$$

After few calculations (4.33) is written

$$x_1 = x + C_{20}\left(x+\frac{C_{11}}{2C_{20}}y\right)^2 + \left(C_{02}-\frac{C_{11}^2}{4C_{20}}\right)\left[y-\frac{2C_{20}^2}{C_{11}}\left(x+\frac{C_{11}}{2C_{20}}y\right)^2\right]^2$$

$$y_1 = y - \frac{2C_{20}^2}{C_{11}}\left(x+\frac{C_{11}}{2C_{20}}y\right)^2 + \left(D_{20}+\frac{2C_{20}^2}{C_{11}}\right)x_1^2 \tag{4.35}$$

whose second degree terms are the same as those of T_2.

The system of equations (4.34) has always a solution except in the case where C_{20} or C_{11} vanishes not simultaneously. This case can be solved using the above method if one of the following changes of variables :

$$x = X \;,\; y = X-dX \;; \tag{4.36}$$

or

$$x = X-dY \;,\; y=Y \tag{4.36'}$$

is first performed on (4.31). In fact, the new C_{ij}, D_{ij} coefficients obtained by means of this change of variables are :

$$C'_{20} = C_{20} + dC_{11} + d^2C_{02} \;,\; C'_{11} = C_{11} + 2dC_{02} \quad \text{with (4.36)} \tag{4.37}$$

$$C'_{20} = C_{20} - dD_{20} \;,\; C'_{11} = C_{11} + 2dC_{20} - dC_{11} - 2d^2C_{20} \quad \text{with (4.36')} \tag{4.37'}$$

In order not to destroy the conservative nature of the problem it is necessary for this change of variables to have constant jacobian. This is obviously the case. Practically it is better to use (4.36) when $C_{20}= 0$ and (4.36') when $C_{11}= 0$.

At this stage the following remark can usefully be made. If we take the expressions (4.35) it can be seen that the terms which are of degree higher than two have coefficients depending on the ratio $C_{11}/2C_{20}$ and its inverse. If one of these forms has an absolute value much greater than one, it may happen that the term in question increases considerably even in a small neighbourhood of the origin, so that the approximating mapping may fail to give a good idea of the behaviour of solutions except in the most immediate vicinity of the origin This means that the parameter d should be chosen in order to increase the size of the neighbourhood in which exact and approximate transformations have similar behaviour and in the cases where it is not necessary to perform osculation (C_{20} and C_{11} non null) it may be regarded as an additional degree of freedom (see the examples at the end of the chapter).

IV.2.6. Approximating Mapping. Osculation Up To Third Degree Terms

Let us now consider the third degree polynomial mapping T_2:

$$x_1 = x + \sum_{m,n} C_{mn}\, x^m y^n \qquad\qquad m+n=2,3 \qquad (4.38)$$
$$y_1 = y + \sum_{m,n} D_{mn}\, x^m y^n$$

assuming that the transformation T_1 has been already made. The osculation method is very similar to that for the second degree, although there are certain minor difficulties which are discussed below.

First from the area-preserving property of the full mapping M one gets the following relations if the jacobian is written as being equal to 1.

$$D_{11}+ 2C_{20} = 0$$
$$2D_{02}+ C_{11}= 0$$
$$D_{21}+ 2C_{20}D_{11}+ 2C_{30}- 2C_{11}D_{20}= 0 \qquad (4.39)$$
$$2D_{12}+ 4C_{20}D_{02}+ C_{11}D_{11}+ 2C_{21}- 4C_{02}D_{20}= 0$$
$$3D_{03}+ 2C_{11}D_{02}+ C_{12}- 2C_{02}D_{11}= 0$$

Thus, of the 14 coefficients of (4.38), at most nine are independent. Using the same procedure as in the previous section, we begin with a third degree polynomial of the form :

$$x'= g(x,y) = x+\beta[(\alpha x+\beta y)^2+ a(\alpha x+\beta y)^3]$$
$$y'= h(x,y) = y-\alpha[(\alpha x+\beta y)^2+ a(\alpha x+\beta y)^3] \qquad (4.40)$$

which is area preserving, then we construct the mapping (still area preserving)

$$x_1= x'+\delta y'^2+\varepsilon y'^3= g(x,y)+\delta h(x,y)^2+\varepsilon h(x,y)^3$$
$$y_1= y'+\gamma x_1^2 +\nu x_1^3 = h(x,y)+\gamma x_1^2+\nu x_1^3 \qquad (4.41)$$

The resulting mapping $(x,y) \longmapsto (x_1,y_1)$ contains seven independent parameters so that, in the general case, it is not possible to carry out to osculation between (4.38) and (4.40 - 4.41) directly. It was shown above that the parameters $\alpha,\beta,\gamma,\delta$ are sufficient to solve the osculation on the second degree terms. All that is needed therefore is to free two extra parameters which would act on the third degree terms. This can be done by performing on (4.38) a change of variables depending on two parameters. The linear changes (4.36) and (4.36') with parameters d_1,d_2 could be used, but this way is rather cumbersome since such a change would involve the resolution of third order equations with respect to d_1, d_2 to fix them. Furthermore if d_1 or d_2 is used for this purpose, the "degree of freedom" pointed out at the end of section IV.2.5 is lost.

In fact, two transformations of variables are used, chosen not only for their simplicity, but also because the inverse transformations is easily obtainable. Recall that these transformations must have a constant jacobian in order not to destroy the area preserving property of the full mapping. This is obviously the case for

$$T_{t_1} : U=x\ ,\ V=y+t_1x^2 \qquad (4.42)$$

and

$$T_{t_2} : U'=U+t_2V^2,\ V'=V \qquad (4.43)$$

The diagram below indicates the sequence of operations which are performed (full lines)

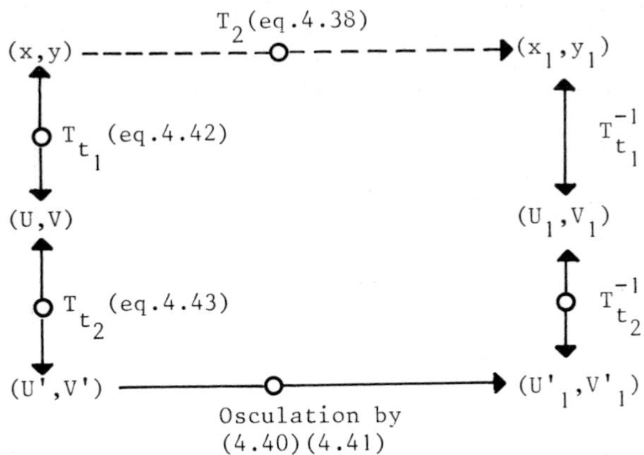

Some details of the calculations for determining the coefficients $a,\varepsilon,\nu,t_1,t_2$ are given in the appendix. The coefficients $\alpha,\beta,\delta,\gamma$ are determined as in the preceding section. It should be noted that the equations obtained are linear with respect to the unknown parameters. For the third order terms the results are

$$\begin{aligned}
t_1 &= [2(D_{11}-C_{20})a_1-2D_{20}b_1]\Delta^{-1}\\
t_2 &= -[3C_{11}r^2a_1-b_1(3C_{11}r-2C_{02})]\Delta^{-1}\\
a &= C_{30}-t_1C_{11}\\
\varepsilon &= C_{03}+(2C_{02}-C_{11})t_2+2C_{02}r\Delta-ar^3\\
\nu &= D_{30}+(2C_{20}-D_{11})t_1-2\gamma C_{20}+ar^{-1}
\end{aligned} \qquad (4.44)$$

where

$$r = C_{11}/2C_{20}\ ;\ \Delta = 2(D_{11}-C_{20})(3C_{11}r-2C_{02})-6D_{20}C_{11}r^2$$

$$a_1 = 3C_{30}r-2\Delta C_{20}r^{-1}-C_{21}\ ;\ b_1=3C_{30}r^2-4\Delta C_{20}-C_{12}$$

Remark : a necessary condition on defining the above parameters is that at least one of the second degree terms of (4.38) should be non null. If there are no second degree terms in (4.38) the changes (4.42) and (4.43) cannot be effective on the third degree terms. What happens in such a case ?

The mapping T_2 has five independent parameters. The mapping

$$\begin{aligned}
x' &= x+\beta(\alpha x+\beta y)^3\\
y' &= y-\alpha(\alpha x+\beta y)^3\\
x_1 &= x'+\varepsilon y'^3,\quad y_1 = y'+\nu x_1^3
\end{aligned} \qquad (4.45)$$

possesses four independent parameters, so, here too, a direct osculation is not available. In this case one can use the linear change (4.36) or (4.36'), but the previous remark is still valid, i.e., the choice of the parameter d will be made by solving a third order equation.

There is another solution which has the advantage of making it possible to use the above results. It consists of introducing second order terms into the mapping T_2 (artificially), obtaining a mapping $\bar{T}_2$ for which the previous calculations can be used, and finally eliminating these second degree terms in the osculating transformation. The choice of these terms cannot be made arbitrarily, since they must be such that $\bar{T}_2$ can be taken as the truncation of an area preserving transformation. This is true for

$$\bar{T}_2 \quad \begin{aligned} x_1 &= x + cy^2 + \sum_{mn} C_{mn} x^m y^n \\ y_1 &= y + \sum_{mn} D_{mn} x^m y^n \end{aligned} \tag{4.46}$$

It is easy to see that, by applying, to the mapping which approximates $\bar{T}_2$, the area preserving transformation

$$\begin{aligned} X &= x - cy^2 \\ Y &= y \end{aligned} \tag{4.47}$$

The consequence is to cancel the term cy^2 in the final mapping which is still area preserving because it is the product of area preserving mappings.

IV.2.7. Example

Let us return to the example of section IV.1.7 with the differential equation

$$\frac{d^2x}{dt^2} = f(t)\, \alpha\, x^2$$

The results obtained using the osculation method have been compared with those obtained by numerical integration (fourth order Runge Kutta method with adaptative step) with double precision (precision of about 10^{-16}). From amongst the many tests carried out, two have been selected as being representative of certain characteristic phenomena of area preserving mappings : island structure, stochastic behaviour. In (54) others examples and other test can be found. Table 1 gives the value of parameters chosen for the osculation in two tests labelled A and B. It should be noted that in this case $r \cong 1$ was found empirically to be a good value for the ratio $C_{20}/2C_{11}$.

A) $T = 1.588$; $\theta = 0.36$; $\alpha = 10$

The results obtained by numerical integration are shown in Fig. 4.6. A brief description of this figure is given below.

A 4th order island structure surrounds the origin which is a center. The regular behaviour of the trajectories breaks down at some limit near the curve Γ. Inside it we have an apparently integrable domain, which will be referred to as a "quasi-integrable" region, since actual integrability cannot be generally verified (Ref. 55). Outside Γ the iterated points display a very irregular motion which we refer to as "stochastic" motion. This behaviour occurs along with homoclinic and heteroclinic points, generated by invariant curves passing through singular saddle points (Ref. 56). In the present example, this type of motion gives rise to stochastic instability (57). Thus outside Γ and with an initial condition near it, the point at first moves slowly diverging with respect to Γ and finally escapes irreversibly from the frame of Fig. 4.6.

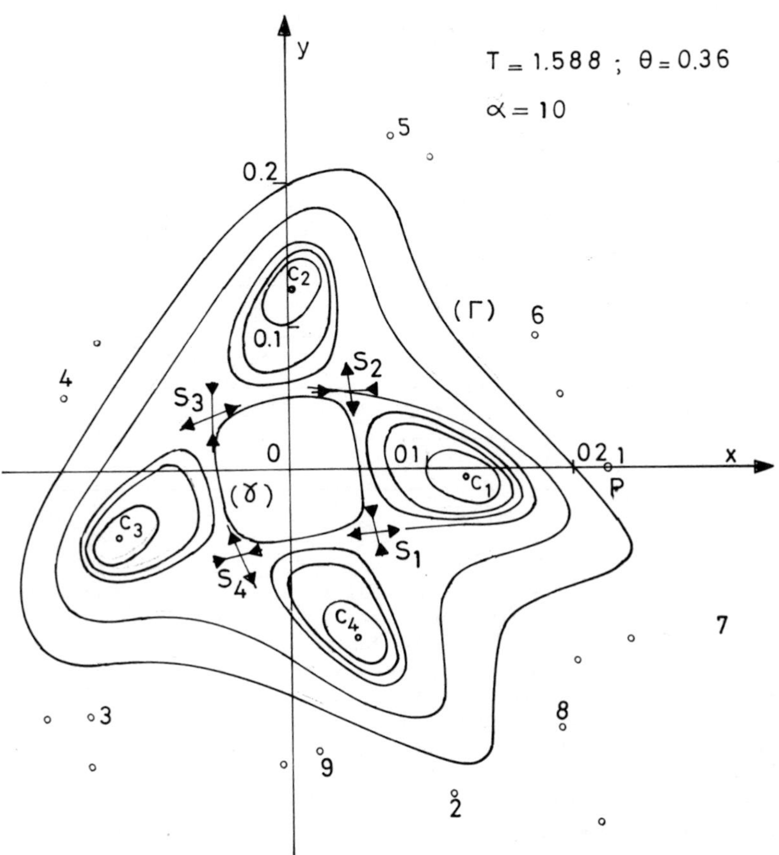
y
T = 1.588 ; θ = 0.36
α = 10
x
0
0.2
0.1
01
02
(Γ)
(γ)
C1
C2
C3
C4
S1
S2
S3
S4
P
1
2
3
4
5
6
7
8
9

Fig. 4.6

The scattered points shown belong to the same trajectory beginning at P (the numbers indicate some successive iterates of the fifth power of the mapping).

TABLE 1

Example / Parameter	A	B
μ	1.150	1.10
C_{20}	3.37347	3.37698
r	0.99912	1.00746
γ	0.03744	0.03705
δ	0.03910	0.03954
t_1	0.01549	0.01551
t_2	0.01657	0.01697
ν	0.00176	0.00173
ε	0.00146	0.00150
a_3	1.38967	1.39126

As regards the comparison between numerical integration and osculation mapping table 2 gives some iterated points obtained by both methods, n denoting the number of iterations, starting from the same initial conditions. Obviously, the divergence between the results increases as the distance from the origin becomes larger. However, a point which should be noted is that in the quasi-integrable region the error occurs in the "phase" rather in the "amplitude". This is reflected in the fact that the phase portrait is almost the same, at least within the degree of accuracy of Fig. 4.6, which therefore also represents the results obtained by the osculating transformation.

The position of the cycles C_1,C_2,C_3,C_4 and S_1,S_2,S_3,S_4 of osculating transformation have been obtained by a numerical program using both the gradient method and Newton's method with a precision of about 10^{-9}. A test using numerical integration shows that the exact and approximate cycle positions agree to a precision of 10^{-5}. Both numerical integration and osculating mapping give invariant curves passing through the saddle S_1,S_2,S_3,S_4 with no detectable homoclinic or heteroclinic points (two of these curves were shown if Fig. 4.6 between S_1 and S_2). This fact is related to the quasi-integrable character of the region inside Γ which contains the island structure corresponding to these saddles.

B) $T.= 1.63$; $\theta = 0.36$; $\alpha = 10$

In this case the integration and the osculation method also agree (see table 3). In figure 4.7 one can see that the island structure is further away from the origin. Around the fourth order center cycles, the behaviour of the solu-

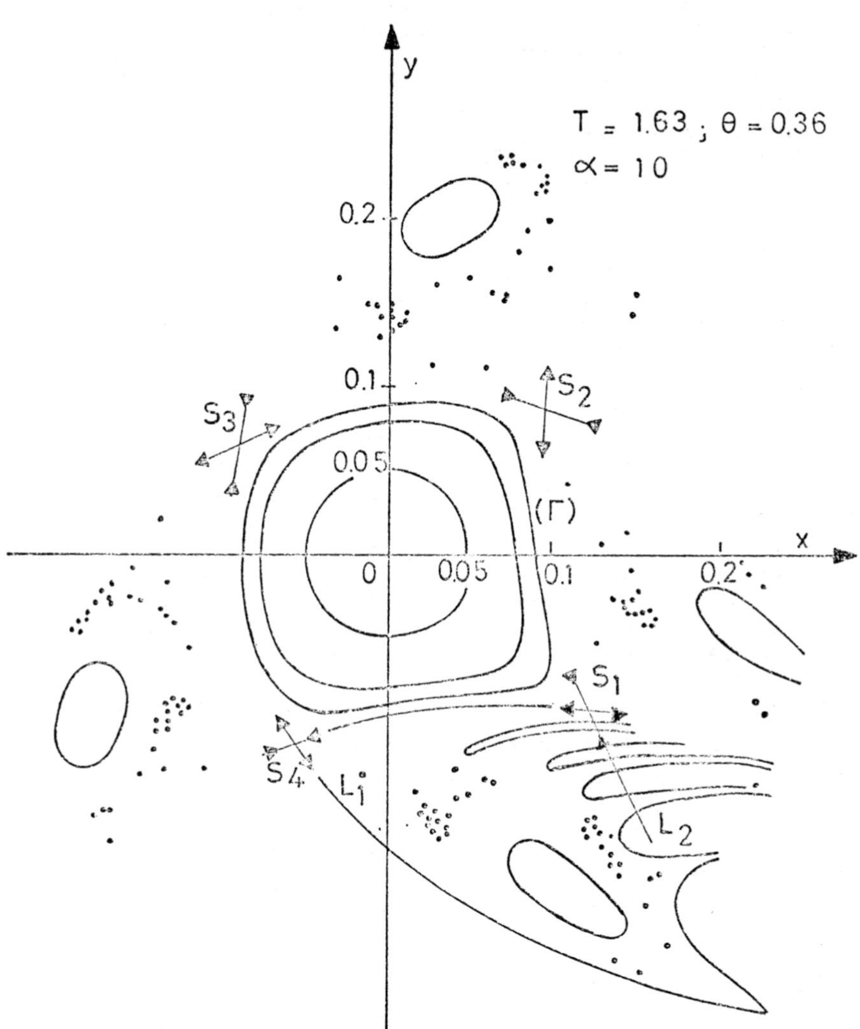

Fig. 4.7

	Osculating mapping		Numerical integration		
n	X	Y	X	Y	
0	0.05	0	0.05	0	
20	-0.00835113	-0.05139432	-0.00834973	-0.05139527	
40	-0.01862271	-0.05294405	-0.01861879	-0.05294774	
60	-0.02587861	-0.05323888	-0.02587089	-0.05324836	
200	-0.05009292	0.00901935	-0.05009050	0.00892006	
400	0.02743005	0.05065437	0.02729490	0.05067465	
600	0.05500973	-0.03379491	0.05503413	-0.03366572	
745	-0.05016173	0.01241045	-0.05016082	0.01204827	
0	0.1	0	0.1	0	
100	0.04383988	-0.09528784	0.04374121	-0.09524857	
200	0.05865594	-0.10588185	0.05846941	-0.10566201	
400	0.06106618	-0.13961848	0.06167375	-0.13966534	
700	0.02937415	-0.09562585	0.02899163	-0.09585284	
0	0.2	0	0.2	0	
100	0.152466	-0.110926	0.15334668	-0.10937630	
200	0.171026	-0.0938892	0.17440640	-0.091378	
400	0.2272165	-0.0720555	0.236263	-0.065829	
700	0.158958	-0.045156	0.140034	0.06575	
0	0.225	0	0.225	0	
10	-0.138654	-0.173515	-0.137934	-0.173137	
20	0.080260	0.233820	0.080038	0.233382	Random zone
30	0.301220	-0.114166	0.301447	-0.113954	
40	0.014320	-0.195143	0.019999	-0.196991	
50	-0.114962	0.107698	-0.130526	0.091030	
0	0.0053833	0.1263964	0.0053833	0.1263694	
8			0.0053750	0.1263506	
16	id.		0.0053735	0.1263323	Center (point C_2)
32			0.0053904	0.1263137	
64			0.005434	0.12636651	
0	0.0447153	0.0525959	0.0447153	0.0525959	
4			0.0447139	0.0525977	
8			0.0447125	0.0525997	Saddle (point S_2)
12			0.0447113	0.0526018	
24			0.044707	0.0526135	

TABLE 2. $T = 1.588$; $\theta = 0.36$; $\alpha = 10$

Example A

tions is regular and defines a four area domain of quasi-integrability. The domain of quasi-integrability around the origin is reduced to the region inside the curve Γ (which is the apparent boundary of this domain) and is separated from the former by a region where the points are scattered. A test of non integrability for this region has been made by drawing some of the analytical invariant curves passing through the points S_1 and S_4 of the fourth order saddle cycle. From S_4, the repulsive invariant curve L_1 does not join the attractive curve L_2 smoothly but intersects it at heteroclinic points. Some loops of L_1 have been drawn on Fig. 4.7 using the osculating mapping, and a verification undertaken by numerical integration shows that at least for the first loops the difference between the results appears to be in the "phase" on these curves as for the regular closed curves of the domain of integrability. This stochastic region is also a region of stochastic instability because the point finally escapes. Similar phase portraits can be seen in (Ref. 48) for example which deals with a quadratic algebraic transformation.

	Osculating mapping		Numerical integration		
n	X	Y	X	Y	
0	0.05	0	0.05	0	
100	0.03727	-0.03919	0.3728	-0.03918	
200	0.05036	-0.00344	0.05035	-0.00342	
400	-0.00726	0.04889	-0.00730	0.04888	
600	-0.04408	-0.02555	-0.04405	-0.02560	
700	-0.00949	-0.04853	-0.00957	-0.04852	
0	0.15	0	0.15	0	
40	0.04454	-0.14284	0.04415	-0.14405	
80	0.04127	-0.15363	0.03822	-0.15158	
120	0.04252	-0.16813	0.03723	-0.16091	Random zone
160	0.04662	-0.18021	0.03290	-0.16470	
200	0.07649	-0.20455	0.02220	-0.15151	
0	0.2	-0.05	0.2	-0.05	
100	0.13841	-0.22519	0.14023	-0.22538	
200	0.13488	-0.19944	0.13655	-0.20198	
300	0.10496	-0.17563	0.10167	-0.17381	
400	0.07614	-0.17172	0.07838	-0.17138	
600	0.13197	-0.22343	0.12287	-0.21994	

TABLE 3. $T=1.63$; $\theta = 0.36$; $\alpha = 10$

Example B

IV.3. Conclusion

If the numerical study of the solutions of a Hamiltonian system with periodic coefficients is undertaken by means of numerical integration, then one comes up against the problem of the sensitivity of the system to the discretisation of the differential equations. This can be explained by the conservative nature of the system under consideration. This problem can, broadly speaking, be solved when it is possible to associate with the system a pointwise transformation. Unfortunately however, it is not usually possible to formulate such a transformation. It is possible to determine its first terms, but the resulting truncated transformation loses its conservative nature. From the truncated pointwise transformation however one can construct an area preserving polynomial pointwise transformation which has, in a sufficiently wide neighbourhood of the origin, the same topological portrait as the exact transformation. In addition, the numerical study is much quicker, since it uses a mathematical model made up of recurrence equations. This last point is an important one since these numerical experiments are at present an important tool in the study of conservative dynamical systems of this kind (Ref. 48,49,57-59).

APPENDIX

From $x_1 = x + \sum_{m+n=2,3} C_{mn} x^m y^n$, $y_1 = y + \sum_{m+n=2,3} D_{mn} x^m y^n$

after having performed the change of variables (4.42) and (4.43) one gets

$$U'_1 = U' + \sum_{m+n=2} C_{mn} x^m y^n + \sum_{m+n=3} C'_{mn} x^m y^n$$

$$V'_1 = V' + \sum_{m+n=2} D_{mn} x^m y^n + \sum_{m+n=3} D'_{mn} x^m y^n \qquad (4.48)$$

where

$$C'_{30} = C_{30} - t_1 C_{11}$$

$$C'_{21} = C_{21} - 2t_1 C_{02} + 2t_2 D_{20}$$

$$C'_{12} = C_{12} + 2t_2 D_{11} - 2t_2 C_{20}$$

$$C'_{03} = C_{03} + 2t_2 D_{02} - t_2 C_{11}$$

$$D'_{30} = D_{30} - t_1 D_{11} + 2t_1 C_{20}$$

$$D'_{21} = D_{21} - 2t_1 D_{02} + 2t_1 C_{20}$$

$$D'_{12} = D_{21} + 2t_1 C_{02} - 2t_2 D_{20}$$

$$D'_{03} = D_{03} - t_2 D_{11}$$

The osculation is performed on the mapping (4.48), by identifying it with the mapping (4.41). The coefficients $\alpha, \beta, \delta, \gamma$ are fixed by the second degree terms which remain unchanged : equation (4.34). By identifying the third degree terms one gets :

$$a\alpha^3\beta = C_{30} - t_1 C_{11}$$

$$3a\alpha^2\beta^2 - 2\beta^2\alpha\delta = C_{21} - 2t_1 C_{02} + 2t_2 D_{20}$$

$$3a\alpha\beta^3 - 4\beta\alpha^2\delta = C_{12} + 2t_2 D_{11} - 2t_2 C_{20} \qquad (4.49)$$

$$a\beta^4 - 2\alpha^3\delta + \varepsilon = C_{03} + 2t_2 D_{02} - t_2 C_{11}$$

$$-a\alpha^4 + 2\beta\alpha^2\gamma + \nu = D_{30} - t_1 D_{11} + 2t_1 C_{20}$$

The first three equations of (4.49) enable one to find the parameters a, t_1, t_2, and finally ε, ν are determined by the last two equations.

CHAPTER V

PRACTICAL FEATURES, NUMERICAL TOOLS — EXAMPLES

This chapter gives an account of some of the practical problems that arise in the analysis of the behaviour of solutions in a second order point mapping

$$x_{n+1} = f(x_n,y_n), \quad y_{n+1} = g(x_n,y_n) \tag{5.1}$$

These problems were first discussed in chapter 1. There, a general survey of the results and their practical use was given. Some initial insight into the behaviour of solutions can be gained by finding the singularities, i.e. the double points and cycles of order k. This will be dealt with in section 5.1. In order to get a more precise idea of the solutions it is often useful to draw invariant curves through saddle points or nodes. This may make it possible to establish, for instance, the domain of influence of some singularities. These two problems deal with a particular mapping of form (5.1), i.e. with fixed values for the parameters and coefficients. Sometimes, a study in the space of parameters and coefficients is needed for practical purposes, either because their values are uncertain or because of slow variations of some parameters. This leads to the analysis of bifurcation cases. Finally some examples are treated.

V.1. Determination of Singularities

The problem is that of determining the double points or k order cycles, of the second order autonomous pointwise transformation (5.1). For reasons of simplicity it will be assumed :

(i) that the non linear functions f and g which appear in the right hand side of (5.1) are analytic in a domain G.

(ii) that the quantities :

$$\begin{aligned} p(x,y) &= x - f(x,y) \\ q(x,y) &= y - g(x,y) \end{aligned} \tag{5.2}$$

have no common factor $h(x,y)$.

In practical terms these are not strong contraints.

Given these assumptions, the pointwise transformation (5.1) has a finite number of singularities in the domain G.

Let ξ be a stable equilibrium set of the pointwise transformation T (5.1), and D its domain of attraction. This set can be obtained by using T to calculate the successive consequents M_n of an initial point M_o belonging to D. Similarly it is possible, using the inverse transformation T^{-1}, starting from an initial condition M'_o, for the succession of iterates to tend towards a stable equilibrium set for T^{-1} (therefore unstable for T).

This "method" of determining singularities, which is no more than a simple simulation method, has a number of major disadvantages e.g.

- it poses the problem of the choice of initial conditions $M_o(M'_o)$.
- certain singularities (e.g. saddle type, which is attractive neither by T, nor by T^{-1}) are not taken into account by such a procedure.

There are other methods of determining singularities which involve the solution of a system of non linear algebraic equations :

$$p(x,y) = q(x,y) = 0 \tag{5.2}$$

- algebraic method, which can only be considered in the simplest cases where f and g are polynomials of low degree (2, 3 maximum).
- graphic method, which involves drawing the curves $p(x,y) = 0$, $q(x,y) = 0$ in the (x,y) plane. This also can only be used in simple cases (e.g. linear piecewise recurrences).(Ref. 60)
- analog method, which consists of associating a dynamic system of the form :

$$\begin{aligned} \frac{dx}{dt} &= P(x,y) \\ \frac{dy}{dt} &= Q(x,y) \end{aligned} \tag{5.3}$$

to the static system (5.2), P and Q being chosen such that :

α) $p(x,y) = q(x,y) = 0 \implies P = Q = 0$
β) the solutions of (5.2) are stable equilibrium positions for (5.3).

Forms such as :

$$\begin{aligned} P(x,y) &= a(x,y)\ p(x,y) + b(x,y)\ q(x,y) \\ Q(x,y) &= c(x,y)\ p(x,y) + d(x,y)\ q(x,y) \end{aligned} \tag{5.4}$$

clearly fulfil condition α, condition β being much more difficult to satisfy. The coefficients a, b, c, d, which correspond to the gradient method and the Newton method (ref. 61), provide a system (5.3) which satisfies condition β but which may also give other stable solutions, i.e. equilibrium positions for (5.3) which are not roots of (5.2).

A derived method which we shall call the integrator method, consists of choosing the coefficients a, b, c, d as constants ($ad - bc \neq 0$).

This necessitates choosing several systems like (5.3), since in general all the equilibrium points are not stable for a given matrix $\begin{bmatrix} a & b \\ c & d \end{bmatrix}$. The initialisation problem also arises for this method. Discretisation of the system (5.3) provides a numerical method; however, this has the disadvantage of requiring a choice not only of initial conditions, but also of rate of discretisation, the latter involving a compromise between speed and accuracy.

V.1.1. Brief Description of the Proposed Method (Ref. 61)

The method is based on the "integrator" method, and makes use of the properties of coarse systems, also called structurally stable systems (Ref. 76).

A detailed account of these systems can be found in the work of Andronov (2).

A brief description is given in section V.1.2.

Let P be a double point of the pointwise transformation (5.1) i.e. a root of the equation (5.2) and S_a the set of differential systems of the type (5.3) (integrator method) whose equilibrium positions are the roots of (5.2). The method is as follows :

1. By a suitable choice of the matrix of the coefficients a b c d, define a system $S_a^+ \in S_a$ for which P is a saddle point. The necessary and sufficient condition for this to be possible is

$$\frac{\partial p}{\partial x}\frac{\partial q}{\partial y} - \frac{\partial p}{\partial y}\frac{\partial q}{\partial x} \neq 0 \ ;$$

This will be seen to be a condition imposed by the assumption that the differential system (5.3) is coarse.

Thus, two principal axes A_1 and A_2 can be defined in P (Fig. 5.1), tangential to two invariant curves L_1 and L_2 which are separatrices of the domains of attraction of stable sets for $T^-[L_1]$ and T^{-1} $[L_2]$ respectively. These sets may be at a finite or an infinite distance.

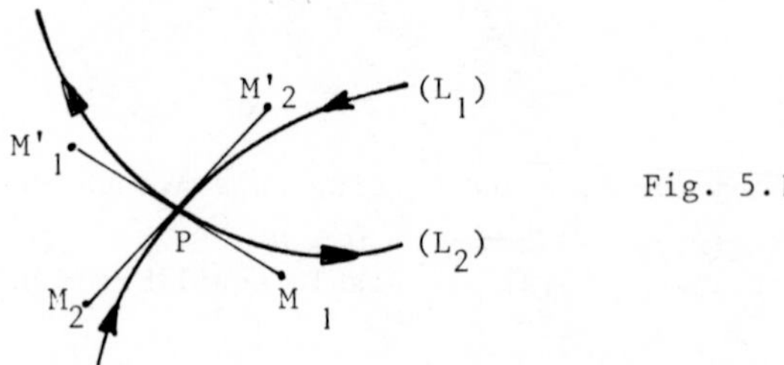

Fig. 5.1

2. The two trajectories of the differential system S_a^+, initialised at t = 0 in M_1 and M'_1, should enable two stable sets to be determined for S_a^+. If we now consider the system S_a^-, obtained from the preceding one by making t = -t, the two trajectories initialised in M_2 and M'_2 should enable two stable sets for S_a^- (unstable for S_a^+) to be determined.

Since the differential systems S_a^+ and S_a^- are autonomous second order differential systems, the limit sets obtained are :
- either equilibrium positions (isolated points) i.e. roots of (5.2)
- or limit cycles, closed curves which are also separatrices and inside of which there may be other limit cycles whereas there must be at least one equilibrium position for S_a^+ (or S_a^-) i.e. one root of (5.2).

At most four new solutions are obtained in this way. For each one of these the above procedure can be reapplied : the choice of a differential system S_a^+... This constitutes a systematic search for the roots of equation (5.2) and solves the problem of determining the initial conditions.

V.1.2. Coarse Systems. Some Results

In the present section a brief description of coarse systems is given. In the light of this, the importance of the role played by points of saddle type in the proposed method can be understood more clearly.

By definition the differential system (5.3) is said to be coarse in a certain domain D, if the topology of the following differential system :

$$\frac{dx}{dt} = P(x,y) + \Delta P(x,y)$$
$$\frac{dy}{dt} = Q(x,y) + \Delta Q(x,y) \qquad (5.5)$$

(where ΔP, ΔQ are algebraic functions with sufficiently small first derivatives) is identical to that of (5.3), in other words, if the phase portraits of the systems (5.3) and (5.5) correspond in a topological tranformation (one to one, continuous).

Before going on to characterise the properties of coarse systems, some definitions are in order.

L_o represents a trajectory passing through the point M_o ; it is made up of two half trajectories : L_o^+ : locus of the points $x(t)$, $y(t)$ when $t \in [0,+\infty[$, $x(t)$, $y(t)$ being the solutions of the system initialised at $t = 0$ by the coordinates of the point M_o.

L_o^-: the half trajectory L_0^+ of the differential system (5.3) in which we made $t = -t$. The half trajectory L_o^+ is said to be orbitally stable (or non singular) if :

$$\forall \varepsilon > 0, \ \exists \delta(\varepsilon) \text{ such that if } d\,[M_1 - M_0] < \delta \rightarrow d[L_1^+ - L_0^+] < \varepsilon$$

where d [.] is some geometrical distance. In the opposite case, the half trajectory L_0^+ is said to be singular. The same definition can obviously be applied also to the other half trajectory L_0^- . Finally, a (full) trajectory L_0 is singular if at least one of its half trajectories is singular.

It can be shown that if P and Q are assumed to be analytic in a domain D , the number of singular trajectories in this domain is finite.

These singular trajectories define zones or cells in the (x,y) plane, i.e. domains of non-singular trajectories with the same behaviour. These cells are finite in number.

The singular trajectories are essentially of three sorts :
- equilibrium points
- limit cycles
- separatrices, i.e. trajectories which leave or which tend to equilibrium points.

The basic properties of coarse systems described by differential equations of type (5.3) can be defined in terms of three conditions
(i) at every equilibrium point the roots of the characteristic equation have a real part which is non-null i.e. there is no equilibrium point which corresponds to a critical case (in the Lyapunov sense).
(ii) the limit cycles are simple
(iii) there is no separatrix which joins one saddle to another.

The following results are essential for the proposed method :
(i) the cells of the plane, (corresponding to identical behaviour of the trajectories within them) are bounded either by limit cycles or by separatrices
(ii) within a limit cycle there is either another limit cycle or at least one equilibrium point
(iii) all the separatrices which are not limit cycles are bounded at one end by an equilibrium point of saddle type, whilst the other end is either bounded by a node or focus, or else it coils around a limit cycle.

It is (iii) which brings out the importance of saddle type points in finding the solutions of the system of algebraic equations (5.2) by associating with it the differential system (5.3) (integrator method).

Knowing all the equilibrium points of saddle type of the differential system (5.3) enables all the equilibrium positions to be determined. To obtain them it is simply necessary to describe all the separatrices starting from the saddles and going as far as the nodes or foci which bound them or the limit cycles around which they are coiled. In this latter case, a simulation can be carried out within the limit cycle until an equilibrium position is determined ; the same procedure can be reapplied if necessary.

V.1.3. Practical Implementation

In this section we give a general outline of the method of finding the solutions of the non-linear algebraic system :

$$p(x,y) = 0 \qquad q(x,y) = 0 \tag{5.6}$$

A more detailed account can by found in (Ref. 61). The solutions of (5.6) are the equilibrium points of the differential system :

$$\begin{bmatrix} \frac{dx}{dt} \\ \frac{dy}{dt} \end{bmatrix} = \begin{bmatrix} a & b \\ c & d \end{bmatrix} \begin{bmatrix} p(x,y) \\ q(x,y) \end{bmatrix} \quad \text{where } ad - bc \neq 0 \tag{5.7}$$

One solution of (5.6) must be known in order to start off the method. The matrix of the coefficients of (5.7) is then determined so as to make this equilibrium point a saddle point. This results in a system : (S^+). A system $(S'+)$ is then chosen such that the equilibrium points of (S^+) of node or focus type are points of saddle type of (S'^+) and vice versa. In this way, any solution of (5.6) is an equilibrium point of saddle type in one of the systems (S^+) or (S'^+). In (Ref. 61) the following two systems are used :

$$(S^+) \begin{cases} \frac{dx}{dt} = p(x,y) \\ \frac{dy}{dt} = q(x,y) \end{cases} \qquad (S'^+) \begin{cases} \frac{dx}{dt} = p(x,y) \\ \frac{dy}{dt} = -q(x,y) \end{cases}$$

These clearly correspond to the conditions imposed.

By plotting the separatrices of the systems (S^+) and (S'^+) alternately we are able to generate, step by step, all the separatrices of the cells of (S^+) and (S'^+) and thus all the equilibrium points of (S^+) and (S'^+) can be determined.

Let P be an equilibrium point of saddle type for (S^+) ; the separatrices L_1 and L_2 are obtained by integration of the equations of the system (S^+) : it should be noted that these curves are attractives for the trajectories arising from points situated in their neighbourhood. The curves L_3 and L_4 are obtained from the system (S^-) and have, in this system, the same properties as L_1 and L_2 in the system (S^+). Plotting these separatrices gives therefore either other equilibrium positions or limit cycles which can be detected by carrying out tests on the velocity vector $(\frac{dx}{dt}, \frac{dy}{dt})$ by counting for example the number of revolutions it makes (Ref. 61).

Remarks :

1) The application of the above method using a computer requires the various differential systems to be discretised. It should be noted however that since it is not essential to obtain very accurate solutions of the differential

systems, it is not necessary to use a very exact discretisation method. Indeed, the discretisation may itself be adapted to ensure the greatest possible speed (Ref. 61) . Finally, in the neighbourhood of the equilibrium points, Newton's method could be usefully employed to increase the speed of convergence of the method.

2. The above method, which has been proposed to deal with the specific problem of the determination of the double points of the pointwise transformation, can be readily extended to other types of problem, e.g. all those which can be reduced to the search for the solutions of a non linear algebraic system with two unknowns (equation 5.2) :
- determination of m order cycles

$$\begin{aligned} x - f^m(x,y) &= 0 \\ y - g^m(x,y) &= 0 \end{aligned} \qquad (5.8)$$

f^m, g^m represent the right hand side of the expressions which define the mth power of the transformation T (in fact the analytic expressions are not necessary since only the numerical values of (5.8) are useful).
-determination of the antecedents of a point $M_0(x_o,y_o)$, i.e. coordinates (x,y) satisfying

$$\begin{aligned} x_0 - f(x,y) &= 0 \\ y_0 - g(x,y) &= 0 \end{aligned} \qquad (5.9)$$

In Ref. 23 , another application is given, i.e. the search for extrema in two variable criteria with equality and inequality constraints.

V.2. Invariant Curves

In the preceding section, we showed the importance of the role played by the equilibrium points of saddle type and the invariant curves which tend to them, in determining the global topological portrait of the trajectories of a second order, autonomous differential system. The behaviour of the trajectories can be entirely characterised by the equilibrium points, the limit cycles and the invariant curves of the saddles, whilst the invariant curves which are not limit trajectories are involved in the determination of the domains of influence of the various equilibrium states (separatrices).

The situation is more complex in the case of pointwise transformations. Some of the figures of Chapter 4, in particular Figure 4.7, give an idea of this complexity. It is true that these figures relate to conservative systems, obviously not coarse in the Andronov sense. However, this type of behaviour has been observed for certain types of pointwise transformation which are not area preserving. Such behaviour can be broadly labelled coarse since it is relatively independent of small variations in the parameters of the equations being dealt with (Ref. 62) .

In Chapter 3 we described a method enabling a formal autonomous differential system to be associated with a second order pointwise transformation. It was said that in general (Ref. 40) the series which appear in the right hand side of this differential system are divergent, which is why it is called a formal associated differential system. The fact that it is in general impossible to associate with a pointwise transformation, analytic in a domain (D), a differential system which is itself analytic gives some explanation for the diversity of behaviour of the solutions of an autonomous second order pointwise transformation compared with that of differential equations of the same type (autonomous, second order).

Aside from the solutions whose behaviour can be described as stochastic, characterised by the existence of homoclinic and heteroclinic points, and of cycles of any order all unstable in the domain considered (Chapter 4), the behaviour which is new compared to that of differential equations is the existence of equilibrium points of types 2 and 3 (saddles, nodes). In addition in the case of non-unique inverse recurrences, the domains of influence of equilibrium points (fixed point or cycles) can be non connected.

In order to determine the domains of influence D of a fixed point or cycle of a recurrence, one possibility is to use functions of Lyapunov's second method, extended to equations of the type (5.1) (Ref. 8)

Using this indirect method, which gives sufficient conditions, it is possible to obtain an estimate of the domain of influence D, this estimate becoming more accurate only as the complexity of the function V is increased. The direct method, which requires the boundary (F) of the domain D to be drawn in the (x,y) plane, can only be considered in cases where F has a simple structure, i.e. when it is in the form of a continuous, closed line.

It is possible (Ref. 62) to find examples in which stability boundary of an attractive equilibrium set (double point or cycle) is in a more complex form, e.g. an annular zone whose width is non null. This will be termed the stochastic stability boundary. This term refers to the fact that the succession of iterates of a point taken as initial condition of the pointwise transformation in this zone either converges towards the equilibrium state ordiverges from it, and that it is almost impossible to predict a priori what the type of behaviour will be. This type of boundary has been discussed in some detail in (Ref. 62) , on a pointwise transformation obtained from an area preserving recurrence slightly disturbed by a non linear term, the resulting transformation being no longer conservative. There is almost no doubt that the stochastic boundaries which have been observed remain after the disturbance, thus providing evidence of the stochastic behaviour of the solutions of the conservative transformation. If some invariant curves of saddle type points were drawn in this zone, the oscillation, described in Chapter IV, would be observed, with the generation of homoclinic and heteroclinic points.

If we restrict ourselves to the simplest case where (F) consists of a continuous closed curve then the following properties can be stated : since (F) is invariant by T and T^{-1}, it is possible to find on (F)

- unstable double points, unstable cycles of finite or infinite order
- the antecedents of these points

In addition, if a saddle type point belongs to (F), it can only be of type 1, or 2 (the positive multiplier is greater than 1) except perhaps at infinity. (F) therefore consists of the invariant curve tangential to the principal axis associated with the smallest multiplier (Figure 5.2).

This introduction shows that in some cases it may be useful to draw certain invariant curves passing through the saddle type points ; the analyticity of these curves is given in Chapter I.

This problem is dealt with in the following two sections in which an analytic method and a practical numerical method are given.

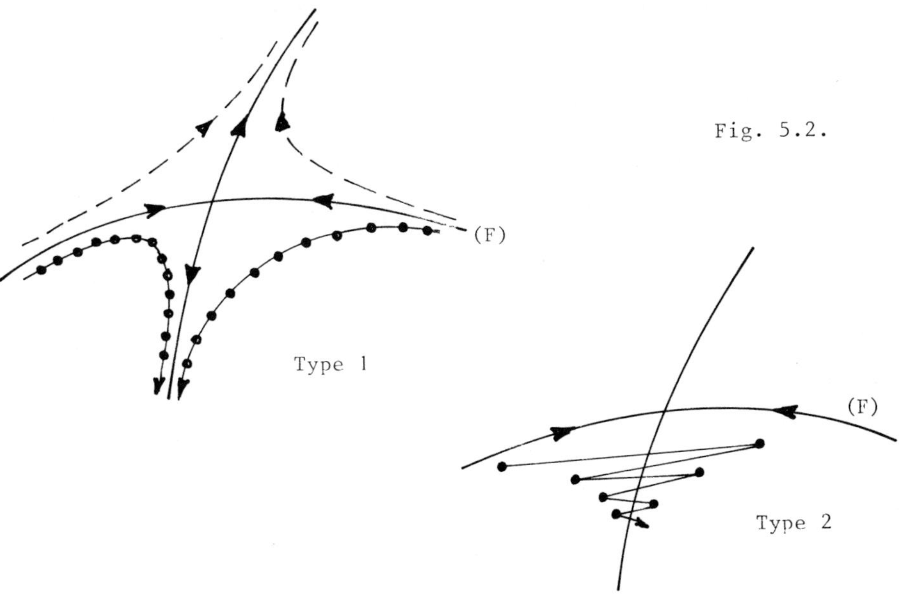

Fig. 5.2.

Remark :
Even in the stochastic boundary case, drawing such lines may constitute a test for complex behaviour by showing up oscillations and the existence of homoclinic points.

V.2.1. Analytic Method
For the sake of simplicity it will be assumed that the origin is a double point of saddle type. This assumption entails no loss of generality. The equations are written in canonical form :

$$u_{n+1} = s_1 u_n + F(u_n, v_n)$$
$$v_{n+1} = s_2 v_n + G(u_n, v_n) \tag{5.10}$$

The axes u, v are the principal axes associated with the eigenvalues s_1 and s_2. In Chapter I we mentioned briefly a method of determining invariant curves tangential to these two axes. The method was developed by Lattes and it consists of writing the invariance directly in the form $v = \theta\ (u)$ (or $u = \theta\ (v)$) Θ being defined from a Taylor series expansion.

This is a direct, rather cumbersome method which in practice allows only the very first terms of the series expansion of θ to be determined. The method described below is due to Picard (63) . It can be programmed numerically, and is therefore somewhat more practical. It is based on the definition of the invariant curve in parametric from :

$$u = f(t),\ v = g(t) \qquad f(o) = g(o) = 0 \tag{5.11}$$

(5.11) is invariant by (5.10) if whenever t is small there is a τ such that :

$$f(\tau) = s_1 f(t) + F(f(t), g(t))$$
$$g(\tau) = s_2 g(t) + G(f(t), g(t)) \qquad (5.12)$$

The problem remains therefore to determine the series f, g and $\tau = \phi(t)$ satisfying (5.12).

If these series converge whenever t is small the curve defined by (5.11) is then analytic. Defining the problem in this way shows that there is one degree of freedom : roughly two equations for three unknowns f, g, ϕ.

The choice (a priori) of the function ϕ should be made so as to simplify the determination of f and g. For example, $\tau = s_1 t$ (or $s_2 t$), (5.12) becomes :

$$f(s_1 t) - s_1 f(t) = F(f(t), g(t))$$
$$g(s_1 t) - s_2 g(t) = G(f(t), g(t)) \qquad (5.13)$$

In practical terms, the direct solution of (5.13) is an insoluble problem. Picard has therefore proposed the following iterative process :

$$f_{p+1}(s_1 t) - s_1 f_{p+1}(t) = F\,[f_p(t), g_p(t)]$$
$$g_{p+1}(s_1 t) - s_2 g_{p+1}(t) = G\,[f_p(t), g_p(t)] \qquad (5.14)$$

He has shown that if f_p and g_p are analytic in the neighbourhood of t = 0, then f_{p+1}, g_{p+1} are also analytic if s_1 and s_2 are distinct real numbers, of absolute values different from 0 and 1, and if $s_2 - s_1{}^k$ $(s_1 - s_2{}^k)$ is different from 0 for any positive integer k. Most important too is the fact that (f_{p+1}, g_{p+1}) tend towards a limit (f, g) when $p \longrightarrow \infty$; f and g are analytic and the solutions of equations (5.13).

Setting

$$f_p(t) = \sum_{p=1}^{\infty} A_p^i t^i \;,\; g_p(t) = \sum_{p=1}^{\infty} B_q^i t^i$$

carrying out the power series expansion in t of F and G :

$$F(f_p(t), g_p(t)) = \sum_{i=2}^{\infty} C_p^i \, t^i$$

$$G(f_p(t), g_p(t)) = \sum_{i=2}^{\infty} D_p^i t^i$$

identifying in (5.14) the terms of the same order with respect to t , we have:

$$A_{p+1}^i = C_p^i\,(s_1^i - s_1)^{-1}$$

$$B_{p+1}^i = D_p^i\,(s_1^i - s_2)^{-1} \qquad (5.15)$$

$$A_{p+1}^1 = \alpha_{p+1} \text{ (arbitrary constant)}$$

$$B_{p+1}^1 = 0$$

where C_p^i, D_p^i are a function of the coefficients A_p^j, B_p^j with $j < i$ i.e. coefficients which have been determined from earlier steps in the identification. The first of these are given below :

$p = 1 \quad f_1(t) = g_1(t) = 0$

$p = 2 \quad f_2(s_1 t) - s_1 f_2(t) = F(0,0) = 0$

$\qquad\quad g_2(s_1 t) - s_2 g_2(t) = G(0,0) = 0$

Thus $\quad f_2 = \alpha_2 t, \; g_2 = 0$

$p = 3 \quad f_3(s_1 t) - s_1 f_3(t) = F(\alpha_2 t, 0)$

$\qquad\quad g_2(s_1 t) - s_2 g_2(t) = G(\alpha_2 t, 0)$

Thus

$$f_3 = \alpha_3 t + \frac{C_2^2 t^2}{s_1^2 - s_1}$$

$$g_3 = \frac{D_2^2 t^2}{s_1^2 - s_2}$$

The coefficients α_i which appear at each step of the calculation can be fixed at the initial value taken for α_2 : $\alpha_i = \alpha_2 \; \forall i > 2$. This is of considerable practical importance, since, given this assumption, it is easy to see that :

$$C_i^2 = C_2^2 \, , \; D_i^2 = D_2^2 \qquad \forall \, i > 2$$

and that, more generally :

$$C_p^k = C_p^p \, , \; D_p^k = C_p^p \qquad \forall \, k > p$$

(5.15) is therefore transformed into :

$$A_{p+1}^i = A_p^i$$

$$B_{p+1}^i = B_p^i \qquad \text{for } i \leq p \tag{5.16}$$

$$A_{p+1}^{p+1} = C_p^{p+1} \, (s_1^p - s_1)^{-1} \quad ; \; B_{p+1} = D_p^{p+1} \, (s_1^p - s_2)^{-1}$$

In other words, the functions f_{p+1} and g_{p+1} which are determined at the pth iteration have the same terms up to and including p^{th} order as the functions f_p and g_p. Because of this, the functions f_i, g_i can be calculated in truncated series form of order i with respect to t, with :

$$f_{i+1}(t) = f_i(t) + A_{i+1} \, t^{i+1}$$

$$g_{i+1}(t) = g_i(t) + B_{i+1} \, t^{i+1}$$

The coefficients A_{i+1}, B_{i+1} are determined by identifying in (5.14) the coefficients of the terms of order i+1 with respect to t. This only involves terms of the functions (f,g) which are of order lower than or equal to (i+1), thus justifying the truncation. It enables the method to be adapted for use with a numerical program. In (Ref. 61) a program enabling the series f and g to be calculated to a degree N fixed in advance ($N \leq 120$) is given.

V.2.2. Numerical Method

The method described in the previous section enables the arc of an invariant curve to be determined. The arc can be bounded by evaluating the error entailed by the truncation of the preceding series expansions (Ref. 61). Further determination of the desired invariant curve can be envisaged, taking into account the following practical consideration :

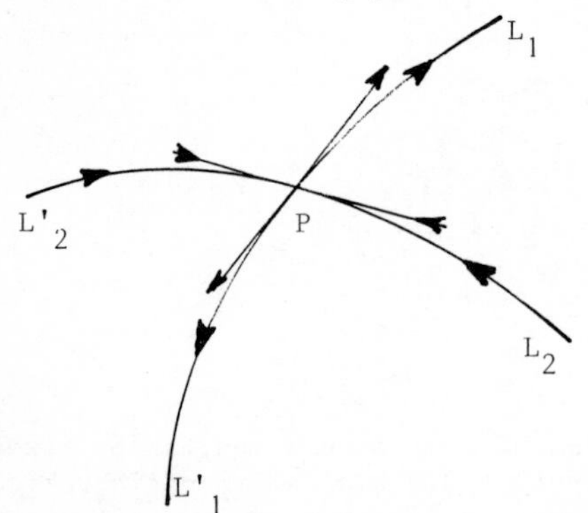

In the neighbourhood of the point P, the invariant curves L_1, L'_1 (L_2, L'_2) are attractive for the successive iterates of the pointwise transformation $T(T^{-1})$. In this neighbourhood, the successive iterates of points taken on the preceding arcs can be considered as belonging to the invariant curves. In most cases, it is even possible by means of a close approximation, to iterate points taken on the principal axes passing through the point P.

The extent of the neighbourhood for which the above is valid cannot be defined exactly. It depends essentially on the type of singularity which is, or ought to be, encountered, when following the various invariant curves from the point P. If for example, the invariant curves tends towards a stable singularity (node, focus, limit cycle) for the pointwise transformation used (T for L_1, L'_1, T^{-1} for L_2, L'_2) then it can be drawn fairly accurately. If on the other hand, the invariant curve enters a zone in which the transformation behaves like a C system (i.e where the behaviour of the tangential pointwise transformation is of saddle type (Ref. 64) which constitutes a test for stochastic behaviour) then the quantitative results become completely false. In fact, in such a zone, the gap between the successive consequents of infinitely neighbouring points increases exponentially. In this case however, the results do have a qualitative value in the sense that they bring out the erratic behaviour of the solutions. The case of conservative pointwise transformations gives examples of such behaviour.

It is implicit in what has been said up to now that, in order to draw the lines L_2, L'_2, the transformation T^{-1} must be known, or the antecedents must at least be easy to calculate. Sometimes this is not so. In that case, a direct method (Ref. 65) can be used : it consists of a systematic search for points with at least one consequent in a given neighbourhood of P defined by the hatched area in the following figure (for example, the point M_2 can be considered as belonging to the invariant curve).

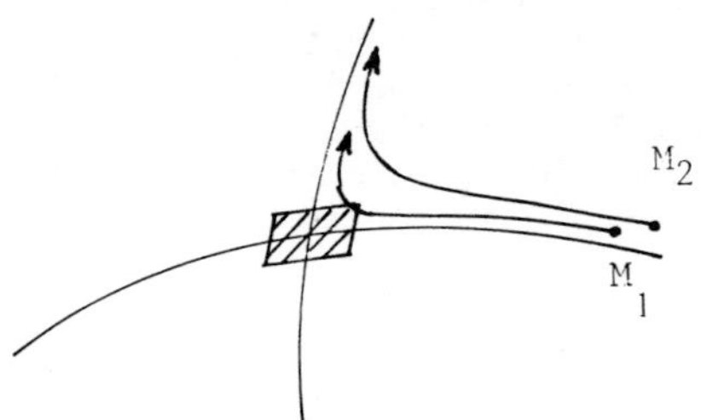

Remarks :

1. In this Chapter we have concentrated on the study of particular solutions of equations (5.1) i.e. double points and k order cycles, invariant curves. To complete the account, another type of equilibrium solution should be mentioned i.e. limit cycles : closed curves, invariant by (5.1) on which there are neither double points nor cycles of any order. As for the case of autonomous, second order differential equations, there are relatively few results available, and few efficient methods of finding such limit cycles. It is easy to formulate a sufficient criterion of non-existence, as Bendixson did for the continuous case :

If in a simply connected domain D of the (x, y) plane, the Jacobian (determinant of the partial derivative matrix) is either greater or lower than 1, then there is no limit cycle in this domain.

The proof is trivial. It involves simply writing the preservation of the area which would be bounded by the limit cycle.

As in the case of differential equations, some work (Ref. 66) has been carried out on certain classes of pointwise transformation in which the non-linear terms depend on a small parameter, in order to determine closed invariant curves (limit cycles).

2. When the inverse transformation T^{-1} is used, it may be useful to determine the critical curves. These are curves which divide the (x,y) plane into zones in which the points have a fixed number of antecedents on each curve the points have two antecedents superimposed. These curves are obtained by taking the consequents of the curves defined by J = 0 (J - Jacobian). They explain the existence of domains of influence of singularities which are not simply connected and which consist of regions with no connexions between them.

V.3. Bifurcation Phenomenon

Consider the second order pointwise transformation :

$$\begin{aligned} x_{n+1} &= f(x_n, y_n, \mu) \\ y_{n+1} &= g(x_n, y_n, \mu) \end{aligned} \tag{5.17}$$

depending on a parameter μ. $\mu = \mu_0$ is said to be a bifurcation value for this parameter if passing through it results in a qualitative change in the behaviour of the solutions of (5.17), i.e. a modification of the topological portrait of these solutions in the x,y plane.

This definition of bifurcation phenomena is very vague and general with the result that an extremely broad range of phenomena can be said to fall under this heading. However, there is one type of bifurcation which has received particular attention, and this is the one which is linked, for a double point (or a k order cycle), to passing through a critical case (in the Lyapunov sense) i.e. when at least one of the multipliers has a value greater than 1. There is no doubt that the interest which has been shown in this type of bifurcation is partly due to the fact that this is the easiest case to tackle : the position of the singular points, the series expansions defining the pointwise transformation in their neighbourhood and the dependence of their coefficients with respect to the parameter, are all known, or assumed known. All that is required, therefore, is a local study before and after bifurcation. It would however be wrong to justify the work which has been carried out in this area purely in terms of its relative simplicity. These bifurcation cases are in fact of more than purely academic interest. If we consider a physical system, operating near the stability boundary (in the parameter space), it is necessary to know the nature of the bifurcation when a certain parameter varies (under the effect of age, for example) in order to ascertain whether the instability brought about by passing through the stability boundary represents a danger, or whether it is a purely local phenomenon which would not prevent the system being used (practical stability). In this context Bautin (11) introduced the notion of dangerous and non-dangerous boundaries.

In general it can be said that :
- if at the moment of bifurcation, the critical case is stable, the change in the behaviour of the solutions will be reversible, the effect of the disturbance of the parameter being small. Reversibility and little effect of parameter change characterize the case of a non-dangerous boundary.
- on the other hand, if the critical case is unstable the reverse situation arises : even very small variations in the parameters can bring about a considerable modification of the solutions (extreme sensitivity). In many cases there is a risk of irreversibility.

Two types of bifurcation are described below :
- bifurcation by passing through the critical case with one multiplier equal to 1.
- bifurcation by passing through the critical case with two complex multipliers $S_{12} = \exp [\pm j \phi]$.

V.3.1. $\underline{S} = +1$

Consider the following first order pointwise transformation depending on a parameter μ, which for the value $\mu = \mu_o$ of this parameter exhibits the critical case under consideration :

$$x_{n+1} = O(\mu) + s(\mu)\, x_n + \sum_{i=2}^{m-1} A_i(\mu)\, x_n^i + A(\mu)\, x_n^m + \dots \qquad (5.18)$$

the dots represent terms of higher order than m, and

$$O(\mu_o) = 0 \ , \ s(\mu_o) = 1, \ \ A_i(\mu_o) = 0 \ \ i = 1, 2, \dots m-1 \ \ , \ A(\mu_o) \neq 0$$

Taking into account the number of independent parameters as well as the independent variations that these parameters may undergo when μ varies, it is easy to see that even in this simple case it is difficult to carry out an exhaustive study of all the possible situations, and especially of the number and nature of the singular points which can appear or disappear when the bifurcation occurs. Nevertheless, it can be seen (Ref. 6) that :

- the number of singular points appearing in the neighourhood of the origin for values of μ close to μ_o depends on the evenness of m : m even, an even number (including 0), m odd, an odd number, therefore always at least 1.
- these points are alternately attractive and repulsive. For $\mu = \mu_o$ they all merge at the origin.

Neimark (67) has made a more detailed study of the particular cases where m = 2 and 3. For m = 2, there may be two neighbouring double points, one stable, the other unstable for μ close to μ_o. Their existence is linked to that of roots for the equation :

$$O(\mu) + (s(\mu)-1)\, x + A(\mu)\, x^2 = 0$$

For m = 3 there may be three double points, roots of :

$$O(\mu) + (s(\mu)-1)\, x + A_2(\mu)\, x^2 + A(\mu)\, x^3 = 0$$

The bifurcation diagram is given in Fig. 5.3, where P is the central double

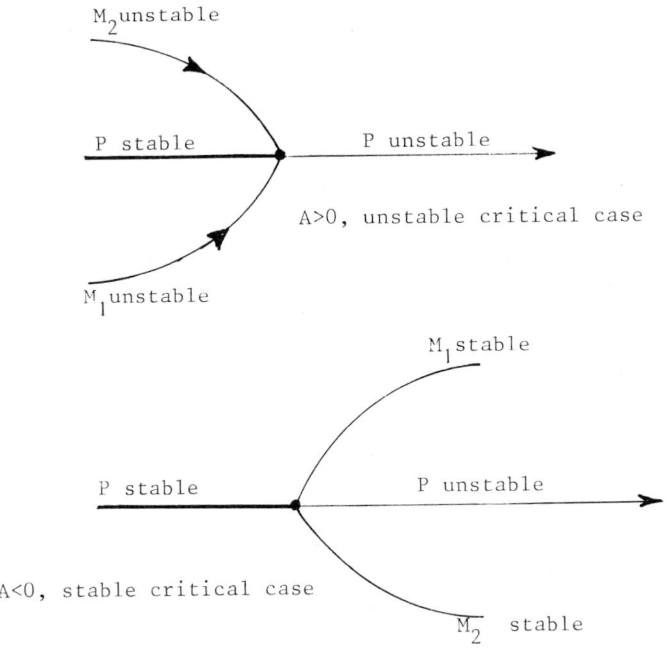

Fig. 5.3

point lying between M_1 and M_2. This bifurcation diagram shows a non-dangerous stability boundary case : by passing through the stable critical case, the central double point, which has become unstable falls between two other infinitely neighbouring double points which are themselves stable.

The case of first order pointwise transformations has been dealt with briefly; however, the study can be extended to transformations of higher order for which only one of the multipliers (associated with a double point or cycle) exceeds the value + 1, under the effect of parameter variation. The extension can be made on the condition that there is, in the neighbourhood of the bifurcation value, a non-singular change of variable (with a non-null Jacobian) enabling the order of the system to be reduced as indicated in Chapter III.

Another type of bifurcation which may be interesting for certain types of application, is that corresponding to a multiplier passing through the value S = -1. This case can be studied in a similar way to the one above by taking the second power of the transformation (5.18). By doing this, it is easy to see that only the m odd case needs to be considered.

In fact :

$$x_{n+2} = O(\mu)+s^2(\mu)x_n + \sum_{p=2}^{m-1} A_i^{(2)}(\mu)x_n^i + A(\mu)(s(\mu) + s^m(\mu) + R(\mu))\, x_n^m + ..$$

where

$$O(\mu_o) = A_i^{(2)}(\mu_o) = R(\mu_o) = 0 \text{ and } s(\mu_o) + s^m(\mu_o) = 0 \text{ for m even.}$$

For m = 3, for the second power of the transformation, the bifurcation cases shown in the Figure 5.3 may be considered.

In the "stable" case, passing through the bifurcation value brings about in the appearance of two double points M_1, M_2 for the second power of the transformation. However, it is easy to check that there are not double points for the transformation (5.18). That is to say, M_1 and M_2 are the two points of a second order cycle, and the bifurcation leads therefore to a doubling of the frequency of the stable cycles (appearance of sub-harmonics). This type of bifurcation has been studied or observed by a number of authors (Ref. 68, 69) and used in an attempt to explain the generation of stochastic solutions (sometimes called chaos) by the accumulation of the bifurcations resulting from passing through the value - 1 as shown below.

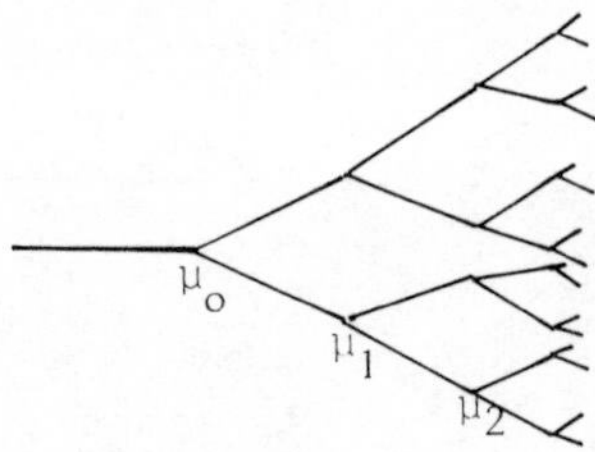

V.3.2. $S_{12} = \exp(\pm j\,w)$

Consider the following pointwise transformation expressed in complex variables :

$$\begin{aligned} x_{n+1} &= r\exp(jw)\,x_n + \sum_{m,n} \bar{C}_{mn}\, x_n^m y_n^n \\ y_{n+1} &= r\exp(-jw)\,y_n + \sum_{m,n} \bar{C}^*_{mn}\, x_n^n y_n^m \end{aligned} \qquad (5.19)$$

The coefficients r, w, $\bar{C}_{mn}$ depend on a parameter μ . Variation in μ causes the multipliers r to exceed the value 1. This type of bifurcation has been studied by various authors (Ref. 71, 72). It is sometimes called Hopf bifurcation (Ref. 73).

There is a non-singular, non linear change of variables, depending continuously on μ (Cigala substitution) which enables (5.19) to be written in the form:

$$\begin{aligned} x_{n+1} &= r\exp(jw)\,x_n + C_{\ell+1,\ell}\, x_n^{\ell+1} y_n^{\ell} + \ldots \\ y_{n+1} &= r\exp(-jw)\,y_n + C^*_{\ell+1,\ell}\, x_n^{\ell} y_n^{\ell+1} + \ldots \end{aligned} \qquad (5.20)$$

in the neighbourhood of the bifurcation value $\mu = \mu_o$ excluding at the same time the exceptional case $w = \frac{2k\pi}{q}$, k, q, are relatively prime.

Specifying the dependence of the various coefficients with respect to μ (5.20) can be written, in polar coordinates :

$$\begin{aligned} r_{n+1}^2 &= R^2(\mu)\, r_n^2 + G_{2\ell+2}(\mu)\, r_n^{2\ell+2} + \ldots \\ \theta_{n+1} &= \theta_n + w(\mu) + \ldots \end{aligned} \qquad (5.21)$$

Finally, it can be seen that this bifurcation case corresponds to that of passing through the value + 1 if we consider the first equation of (5.21). There the equation is simpler, since it has been assumed that the origin remained a fixed point for the transformation (5.19) for all values of μ close to μ_o. The bifurcation is illustrated in Figure 5.4. In the case $\left.\frac{dr}{d\mu}\right]_{\mu_o} > 0$ the direction of the arrow corresponds to that of the increase of μ (the case $\left.\frac{dr}{d\mu}\right]_{\mu_o} < 0$ is obtained by simply reversing the direction of the arrow). The case $\left.\frac{dr}{d\mu}\right]_{\mu=\mu_o} = 0$ is more doubtful, requiring more detailed investigation in particular of the variation of μ around the value $\mu = \mu_o$. This case will not be gone into here.

The above analysis can be extended to the case of m order cycles (m : any integer). This type of bifurcation explains the appearance of solutions which are almost periodic.

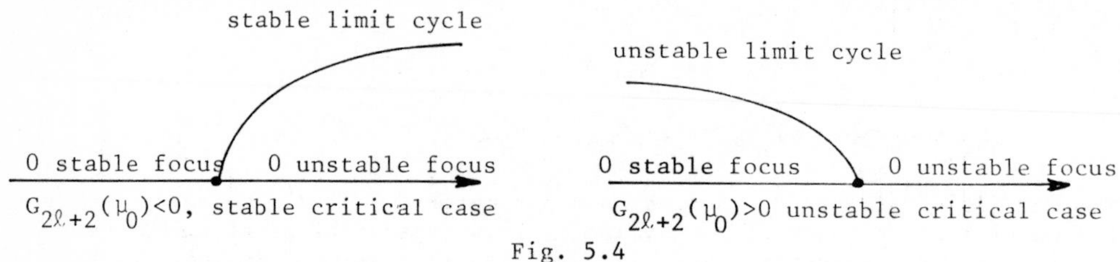

Fig. 5.4

V.4. Some Examples

In this section some examples of the application of the above methods and techniques are given. For the sake of conciseness the systems considered, the method of obtaining the associated recurrent equations and the principal results will be described briefly but no calculation details will be given. These can be found in the works referred to in the text. The first example provides a simple illustration of a bifurcation case. The other belongs to the class of discrete systems which can be described by piecewise, time-invariant, continuous state equations.

V.4.1. First Example

Consider the second order piecewise transformation :

$$\begin{aligned} x_{n+1} &= (1 - \mu)\, x_n + y_n \\ y_{n+1} &= -\, 0{,}2\, \mu\, x_n + y_n - \mu\, x_n^3 \end{aligned} \qquad (5.22)$$

depending on the parameter μ . This transformation will serve to illustrate the case of a non-dangerous stability boundary, for a bifurcation corresponding to one multiplier passing through the value -1. For any μ , the origin is a double point of (5.22), and the value $\mu_o = 4\ (1.8)^{-1}$ is the bifurcation value which causes one of the multipliers at the origin to pass through the value -1, the second multiplier being equal to 0.77.

For $\mu = 2{,}1$ ($\mu < \mu_o$) the origin is a type 2 stable node (Figure 5.5). The numerical program described in (Ref. 61) was used to find cycles for this recurrence. Two fourth order cycles were found :
(i) unstable type 1 node (points A_i, $i = 1, 2, 3, 4$)
(ii) saddle (points B_i), of which certain invariant lines determine the boundary (F) of the stability domain of the origin. This boundary resembles a parallelogram ($\mu = 2$ always results in a parallelogram) and is scarcely affected by an increase in μ after $\mu = 2$. For $\mu = \mu_o$ we have a stable critical case corresponding to one multiplier equal to -1. For $\mu = \mu_o + \varepsilon$ ($\varepsilon > 0$ small) the origin becomes unstable, and a second order cycle appears in the neighbourhood of the origin, M_1, M_2 : stable type 1 node. The boundary (F) enclosing the stability domain of this second order cycle is scarcely affected by this variation, as can be seen in Figure 5.6, for $\mu = 2.3$.

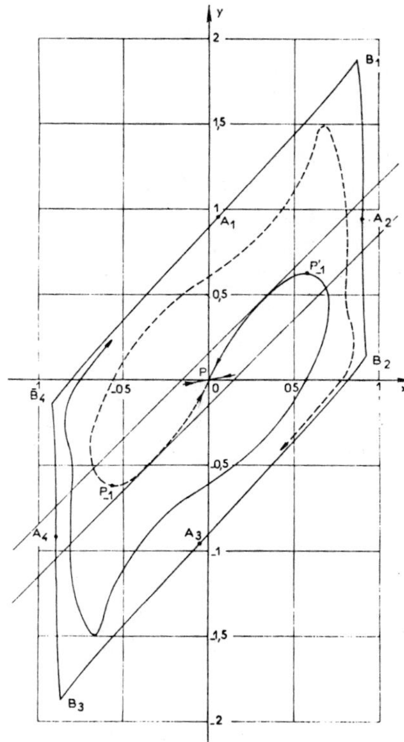
y
2
1,5
1
0,5
-0,5
-1
-1,5
-2
x
-1
-0,5
0
0,5
1
B1
B2
B3
B4
A1
A2
A3
A4
P
P'1
P-1

Fig. 5.5

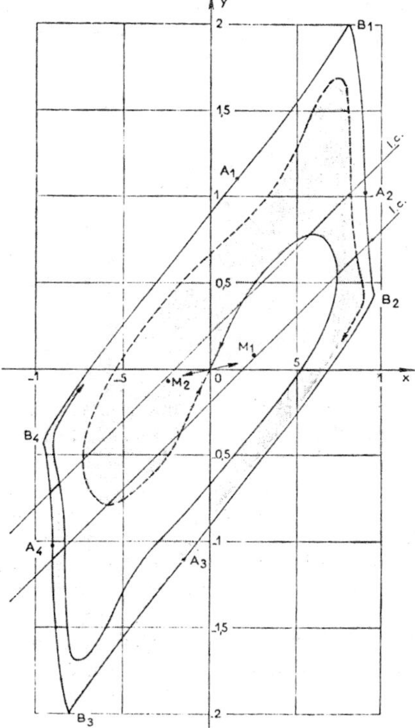
y
2
1,5
1
0,5
-0,5
-1
-1,5
-2
x
-1
-.5
0
.5
1
B1
B2
B3
B4
A1
A2
A3
A4
M1
M2
l.c.
l.c.

Fig. 5.6

V.4.2. Second Example

There are a number of systems particularly in the area of electrical engineering, which can be described by piecewise constant time invariant continuous differential equations written in vector form :

$$\frac{dZ}{dt} = f_i(Z, u_i(t)) \qquad (5.23)$$

where f_i is continuous, and invariant over an interval $[t_i, t_{i+1}]$, t_i, t_{i+1} being the commutation times which may be of different types (Ref.61, 74).

In general, there are three main types of commutation times :
- the fixed commutation times t_f which result from external sampling, usually at a constant rate. In order to maintain a uniform notation, these will be identified using the following equation :

$$g_1(t) = t_f - \ell T - t_o = 0 \qquad \ell : \text{any integer, } t_o : \text{initial time.}$$

- the control commutation times t_c, which correspond to a modification in or to the appearance of a control signal. They are generally modulated by comparing a certain function of the state of the system to a reference signal, usually periodic.

$$g_2(Z(t), A(t)) = 0 \qquad \text{for } t = t_c$$

- the natural commutation times t_N which correspond for example, to the existence in the system of elements which are non linear and discontinuous or with discontinuous derivatives, and are such that :

$$g_3(Z(t)) = 0 \qquad \text{for } t = t_N$$

For the sake of convenience it has been assumed that these commutation times can be defined using a single function g_i, although of course this is not generally the case.

Let (f_1, u_1), (f_2, u_2), (f_3, u_3) be the forms taken by f and u in (5.23) after the times t_f, t_c, t_N respectively and up to the next commutation time which is defined by :

$$\min t : t \in \{ g_1(t) = 0 \text{ or } g_2(t) = 0 \text{ or } g_3(t) = 0 \}$$

The total pointwise transformation which can be associated with (5.23) is given by the product of partial pointwise transformations, corresponding to each determination of the pair (f_i, u_i). These transformations are represented by τ_1, τ_2, τ_3. It is clear that, a priori, the succession of these partial pointwise transformations can be found sequentially in a number of different ways, but if we restrict ourselves to the case of autonomous systems (5.23), this succession will be a function of the initial conditions only . What is more, the total pointwise transformations will be written for the time between two successive commutation times of the same type. This can be explained by the fact that one of the aims of the study is to find the singularities : double points and cycles, which each correspond to a given succession of the various pointwise transformations τ_1, τ_2, τ_3.

Taking a t_f as initial reference time the pointwise transformation which leads to the next t_f can be determined by solving the equations $g_i = 0$,

$i = 1, 2, 3$ sequentially, and thus calculating the τ_i, $i = 1, 2, 3$ successively. (For further details, see (Ref. 61)). The same thing can be done taking a t_c as a reference time. In each case an autonomous recurrence is obtained. A number of problems arises, however, if the t_N are chosen as reference times, since the equation $g_3(Z) = 0$ does not give solutions in t. This creates a problem in particular for the calculation of the Jacobian (partial derivatives matrix), which is necessary for the determination of the singularities and their multipliers.

Whilst the calculation of this partial derivative matrix is theoretically straightforward, it nevertheless poses a number of practical problems. These have been studied for the general case by Rozenvasser (Ref. 75), who proposes the use of a sensitivity matrix, which is established using a step by step procedure. In particular, the calculation involves terms such as $\left.\frac{dt}{dz}\right|_{f,N,c}$, which determine the sensitivity of the commutation times with respect to the variations in the state vector. It is clear that an equation such as $g_3 = 0$ does not allow this since the determination of the sensitivity of the commutation times necessitates knowledge of the previous states of the system. Thus, the values of the system at a time t_N does not constitute a state vector of the system.

As an example let us consider the two phases rectifier of Fig. 5.7 with an inductance-resistance load. The current in the load is compared with a cons-

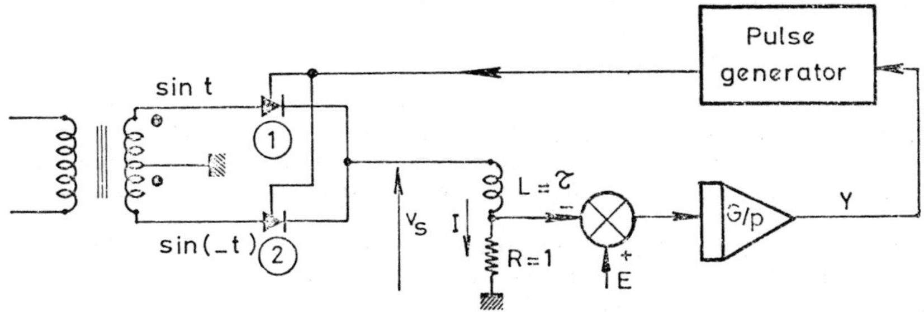

Fig. 5.7

tant reference voltage E giving a control signal by means of a pure integrator and a pulse generator which produces the thyristor firing pulses when the output of the integrator Y reaches the piecewise linear function A(t) given in Fig. 5.8.

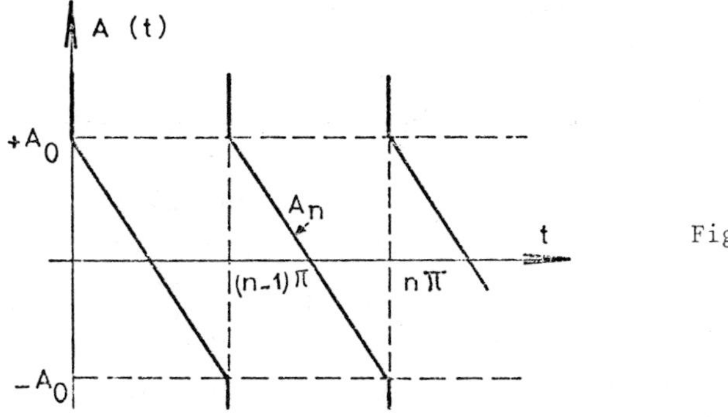

Fig. 5.8

Let $Z^T = [I,Y]^T$ ($[.]^T$ means transpose) be a state vector of the system with normalised variables

a) when a thyristor is conducting, the state equation is :

$$\overset{\circ}{Z} = \begin{bmatrix} -\frac{1}{\tau} & 0 \\ G & 0 \end{bmatrix} Z + \begin{bmatrix} \frac{1}{\tau} & 0 \\ 0 & G \end{bmatrix} \begin{bmatrix} v_s(t) \\ E \end{bmatrix} \qquad (5.24)$$

The function $v_s(t)$ can take the values : $(-1)^{k-1} \sin(t+(k-1)\pi)$, k being odd when the thyristor 1 is conducting and even when 2 is conducting.

b) when no thyristor is conducting the state equation is :

$$\overset{\circ}{Z} = \begin{bmatrix} 0 & 0 \\ -G & 0 \end{bmatrix} Z + \begin{bmatrix} 0 & 0 \\ 0 & G \end{bmatrix} \begin{bmatrix} v_s(t) \\ E \end{bmatrix} \qquad (5.25)$$

and since $v_s(t)$ is not taken into account it can be taken as equal to its previous value. It is clear that there are no t_f times (no external sampling). The commutation times are :

- the time t_c fixed by

$$g_2 = [0,\ 1] \quad Z(t) + A(t) = 0$$

The function f_2 to be considered is the right hand side of (5.24)

-the time t_N fixed by

$$g_3 = [1,\ 0] \quad Z(t) = 0$$

which corresponds to the instinction of a thyristor when the value of the current in the load becomes zero. The function f_3 to be taken is the right hand side of equation (5.25). Figure 5.9 shows in the state space the various cycles obtained for a particular setting of parameters. It should be noted that :

- in addition to the stable first order cycle (which corresponds to the correct behaviour of the system) there exists a high-order stable cycle corresponding to a large amplitude oscillation.

-the other cycles (3rd and 6th order) are unstable and can be used to determine the stability domain of the first order cycle.

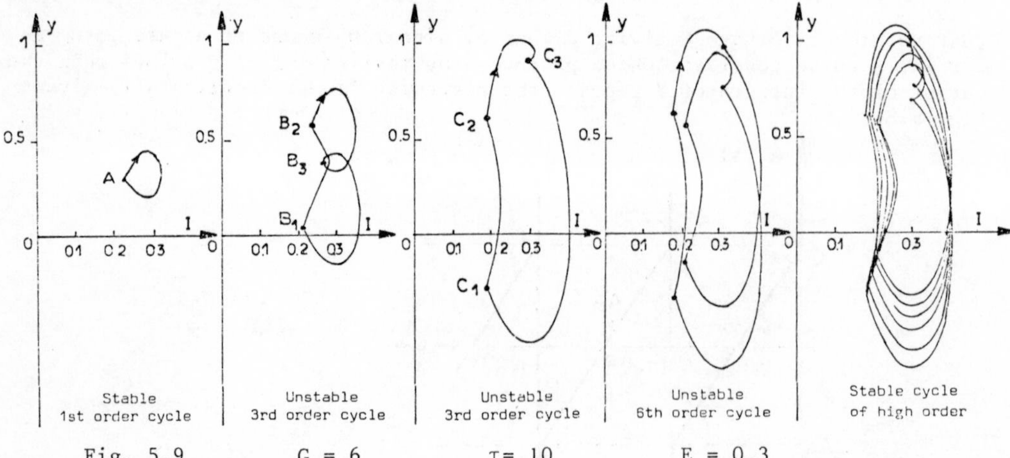

Fig. 5.9 $\quad G = 6 \quad \tau = 10 \quad E = 0,3$

Figure 5.9 represents the results of an analog simulation.the location of the cycles up to sixth order having been previously determined by means of a numerical program (Ref. 61) based on the ideas of section 5.1 and whose results are given in table 5.1.

TABLE 5.1

	CYCLE 1 FOCUS		
A	X = 0,2434 ; SR = - 0,5964 Y = 0,3124 ; SI = 0,6121		
	CYCLE 3 SADDLE		
B_1	X = 0,2173 Y = 0,0573	S1 = 2,2819 S2 = 0,1708	P1 = 10,66 P2 = -1,71
B_2	X = 0,2400 Y = 0,5526	S1 S2 = 0,1708	P1 = -7,05 P2 = 15,58
B_3	X = 0,2670 Y = 0,3790	S1 S2	P1 = 6,80 P2 = 99,80
	CYCLE 3 SADDLE		
C_3	X = 0,3067 Y = 0,9306	S1 = -0,0000 S2 = -1,5218	P1 = 14,44 P2 = -12,62
C_1	X = 0,1973 Y = 0,2544	S1 S2	P1 = 0,68 P2 = 7,22
C_2	X = 0,1983 Y = 0,5919	S1 S2	P1 = -15,93 P2 = - 1,57
	CYCLE 6 SADDLE		
	X = 0,2122 Y = 0,5559	S1 = 0,0000 S2 =-2,0672	P1 = 1,99 P2 = - 4,00
	X = 0,3028 Y = 0,7783	S1 S2	P1 = 27,31 P2 = 14,03
	X = 0,1876 Y = 0,3033	S1 S2	P1 = - 0,16 P2 = 21,11
	X = 0,1885 Y = 0,6158	S1 S2	P1 = -16,80 P2 = 1,71
	X = 0,3026 Y = 0,9695	S1 S2	P1 = 10,28 P2 = - 7,21
	X = 0,2097 Y = 0,1675	S1 S2	P1 = 99,00 P2 = 6,88

SR, SI are the real part and imaginary part of the multiplier, S1, S2 real multipliers, P1, P2 the slopes of the corresponding principal axes. (X stands for I).

The recurrence obtained for these parameters values has two attractive sets. Between these two elements there is a boundary (F) which limits their respective domains of attraction. Among the unstable saddle cycles which are obtained, only one, the third order cycle B_1, B_2, B_3 (which is of type 1) can belong to this boundary since the multipliers of the two others are negative (type 2 saddles). The results obtained by drawing the invariant lines passing through the third order cycle (points B_1, B_2, B_3) are shown roughly in Fig. 5.10. the hatched zone being the domain of attraction of A.

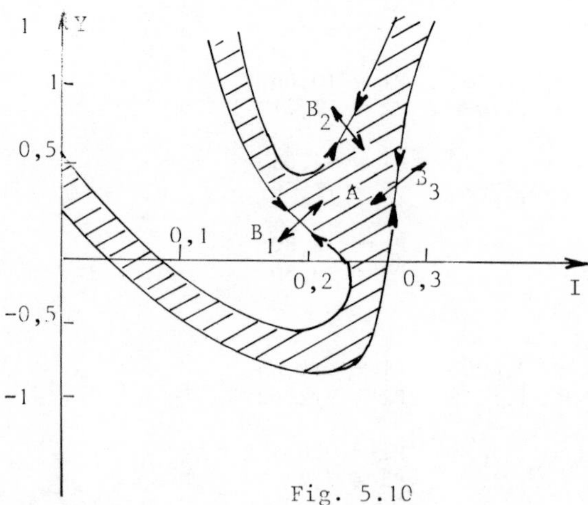

Fig. 5.10

Others examples of application of the method of associated point mapping can be found in(Ref. 21)which analyses the sensitivity to discretisation in numerical integration of differential equations, in (Ref. 6) dealing with non linear sampled data systems and in (Ref. 33) with applications to pulse frequency and pulse width modulated systems.

REFERENCES

1. POINCARE H., Les méthodes nouvelles de la mécanique céleste, Gauthier Villars, Paris, (1892).
2. ANDRONOV A.A., VITT A.A. and KHAIKIN S.E., Theory of oscillators, Pergamon Press, (1966).
3. BLAQUIERE A., Non linear system analysis, Academic Press, (1966).
4. HADAMARD J., Bulletin Société Mathématique Française, Compte Rendu des Séances, (1901).
5. LATTES S., Annali di Mathematica, Serie 3, t. 13, (1906).
6. ROUBELLAT F., Contribution à l'étude des solutions des récurrences non linéaires et application aux systèmes à données échantillonnées, Thèse de Doctorat d'Etat, n° 364, Toulouse, (1969).
7. GUMOWSKI I. and MIRA C., Etude des points singuliers à l'infini d'une récurrence autonome du deuxième ordre à variables réelles, C.R.A.S. Paris, t. 263, p. 547 - 550, Série A, (1966).
8. NEIMARK Yu.I., Izvestia Vuzov Radiofisica, USSR, t. 1, n° 1, (1958), in russian
9. JULIA G., Journal de mathématiques pures et appliquées, 8ème série, 1, p. 47, (1918).
10. FATOU P., Bulletin Société Mathématique Française, t. 47, p. 161, (1919) and t. 48, p. 208, (1920).
11. BAUTIN N.N., Comportement des systèmes dynamiques au voisinage des limites du domaine de stabilité, Goestekhizdat, Moscou, (1949).
12. MALKIN I.G., Theorie der Stabilität einer bewegung, R. Oldenbourg, Munchen, (1959).
13. LYAPUNOV A.M., Problème général de la stabilité du mouvement, Annales de la Faculté des Sciences de Toulouse, 9, p. 204 - 474, (1907).
14. MIRA C., ROUBELLAT F., Etude d'un cas critique pour des récurrences du deuxième ordre, C.R.A.S., t. 266, (1968).
15. LIU HSU, Contribution à l'étude des solutions des recurrences non linéaires Application aux systèmes dynamiques conservatifs. These Doctorat d'Etat, n° 597, Toulouse, (1974).
16. HAHN W., Stability of motion, t. 138, Springer Verlag Heidelberg, New York, (1967).
17. HAHN W., Theory and application of Lyapunov's direct method, Prentice Hall, (1963).
18. ZUBOV V.I., Methods of A.M. Lyapunov and their applications, Noordhoff, Groningen, (1964).
19. OZIRANER A.S., On the stability of motion in critical cases, P.M.M., Vol. 39, n° 3, p. 415 - 421, (1975).
20. BABARY J.P., MIRA C., Sur un cas critique pour une récurrence autonome du deuxième ordre, C.R.A.S., t. 268, Série A, (1969).
21. BABARY J.P., Contribution à l'étude des solutions des récurrences non linéaires et différentes applications, Thèse de Doctorat d'Etat, n° 365, Toulouse, (1969).

22. MIRA C., PUN L., Sur un cas critique d'une récurrence d'ordre m, m > 2 possdédant deux multiplicateurs complexes de module unité, C.R.A.S., Paris, t. 272, Série A, (1971).
23. PUN L., Contribution à l'étude des solutions des récurrences non linéaires. Exemples d'application à différents domaines de l'automatique. Thèse de Doctorat d'Etat, n° 448, Toulouse, (1971).
24. CIGALA A.R., Sopra un criterio di instabilita, Annali di Matematica, Série III, p. 11, (1905).
25. BERNUSSOU J., Sur un cas critique à deux multiplicateurs égaux à un pour une transformation ponctuelle du second ordre, C.R.A.S., Paris, t. 281, (1975).
26. NEMYTSKII V.V., STEPANOV V.V., Qualitative Theory of differential equations, Princeton University Press, New Jersey, (1960).
27. MIRA C., Sur un cas critique correspondant à deux multiplicateurs égaux à l'unité pour une récurrence ou transformation ponctuelle du second ordre, C.R.A.S. Paris, t. 275, (1972).
28. LYAPUNOV A.M., Stability of motion, Academic Press, New York, (1966).
29. DIAMOND P.M., Analytic invariants of mappings of two variables, Jal of mathematical analysis and applications, 27, p. 601 - 608, (1969).
30. BIRKHOFF G.D., Surface transformation and their dynamical applications, Collected mathematical papers, Dover Publications inc, New York, Vol. 2, (1966).
31. LIU HSU, Sur un cas critique à multiplicateurs unitaires pour une récurrence autonome ou une transformation ponctuelle du deuxième ordre à variables réelles, C.R.A.S. Paris, t. 275, Série A, (1972).
32. MIRA C., Cas critique d'une récurrence ou transformation ponctuelle du quatrième ordre à multiplicateurs complexes, C.R.A.S. Paris, t. 272, Série A, (1971).
33. BERNUSSOU J., Contribution à l'étude des solutions des récurrences non linéaires. Application à l'étude de systèmes à modulation, Thèse de Doctorat d'Etat, n° 596, Toulouse, (1974).
34. GUMOWSKI J., MIRA C., Sensitivity problems related to certain bifurcations in non linear recurrence equations, Automatica, Vol. 5, p. 303 - 317, (1969).
35. SIMANOV C.N., KAZEEVA N.I., Diff. Uravnenie, t. 7, n° 5, p. 910 - 918, (1971).
36. KAMENKOV G.V., On the problem of stability of motion in critical case, P.M.M., Vol. 29, n° 6, p. 1053 - 1069, (1965).
37. ARGEMI J., Sur les points singuliers multiples de systèmes dynamiques dans R^2. Ann. di. math. pura e applicada, IV, 79, p. 35 - 70, (1968).
38. BERNUSSOU J., LIU HSU, ABATUT J.L., Sur le cas critique à deux multiplicateurs réels dans une transformation ponctuelle non linéaire du second ordre. Ial Conference on Non Linear Oscillations, Berlin, 8.12 Sept., (1975).
39. BERNUSSOU J., LIU HSU, MIRA C., On non linear periodic differential equation and associated point mapping. Application to stability problems, Ial journal of Non Linear Mechanics,, Vol. 11, p. 1 - 9, (1976).
40. D.C. LEWIS Jr, Formal power series transformations, Duke mathematical journal, Vol. 5, p. 794, (1939).
41. PUN L., Initial conditionned solutions of a second order non linear conservative differential equation with a periodically varying coefficient, Journal of the Franklin Institute, Vol. 295, n° 3, (1973).
42. CROSBIE E.E., KHOE T.K. and LARI R.J., The effects of a delta function sextupole field on phase space trajectories, IEEE Transactions on Nuclear Science, n° 3, p. 1077 - 1081, (1974).

43. LASLETT L.J., Mc MILLAN E.M. and MOSER J., Long term stability for particles orbits, NYO-1480-101, New York University, report.
44. ZLATUSTOV B.A., OKHOTSIMSKII D.E., SARICHEV V.A., TORZEVSKII A.P., Oscillations d'un satellite dans le plan d'une orbite elliptique, Kossm. Issl, n° 2, p. 657, (1964), in russian.
45. MIRA C., Etude d'un premier cas d'exception pour une récurrence ou une transformation ponctuelle, autonome du deuxième ordre à variables réelles, C.R.A.S. Paris, t. 269, Série A, p. 1006 - 1009, (1969).
46. MIRA C., Etude d'un second cas d'exception..., C.R.A.S. Paris, t. 269, Série A, p. 332 - 335, (1970).
47. MIRA C., Sur les cas d'exception..., C.R.A.S. Paris, t. 270, Série A, p. 466 - 469, (1970).
48. HENON M., Numerical study of some quadratic area preserving mapping, Quarterly of applied mathematics, 27, (1969).
49. GUMOWSKI I., Some properties of large amplitude solutions of conservative dynamic systems. Part I, Quadratic and cubic non linearities, CERN/SI/Int. BR/72.1, Genève, (1972); Part II, Bounded non linearities, CERN/SI/Int. BR/72.1, (1973).
50. GUMOWSKI I., MIRA C., Sur la distribution des cycles d'une recurrence ou transformation ponctuelle conservative du deuxième ordre, C.R.A.S. Paris, t. 274, (1971).
51. Mc MILLAN E.M., A problem in the stability of periodic systems, Topics in modern physics, (1970).
52. BERNUSSOU J., LIU HSU, Analysis of conservative systems by approximating area preserving point mapping, Journal of the Franklin Institue, Vol. 299, n° 3, (1975).
53. ENGELS W., Ganze Cremona transformationen von primzahlgrad in der ebene, Math. annalen, Bd 136, p. 319 - 325, (1958).
54. BERNUSSOU J., LIU HSU, Numerical study of periodical hamiltonian systems by means of associated point mapping. Quarterly of applied mathematics, p. 149 - 171, July, (1976).
55. SIEGEL C.L., MOSER J., Lectures on celestial mechanics, Springer Verlag, band 187, (1971).
56. BIRKOFF G.D., SMITH P.A., Structure analysis of surface transformations, Coll. math. papers, Dover Publications inc., Vol. 2, p. 360 - 394 (1968).
57. CHIRIKOV B.V., Research concerning the theory of non linear resonance and stochasticity, C.E.R.N. Translation, 71 - 40, Geneva, (1971).
58. FROESCHLE C., Numerical study of the stochasticity of dynamical systems with more than two degrees of freedom, Journal of Computational Physics, Vol. 11, n° 3, (1973).
59. FORD J., Empirical determination of integrability of non linear oscillators systems using area preserving mappings. Actes du colloque international "Transformations ponctuelles et applications", Toulouse, 10-14 Sept., (1973).
60. GUMOWSKI I., MIRA C., Determination graphique de la frontière de stabilité Electronics Letters, Vol. 2, n° 7, (1966).
61. GIRAUD A., Application des récurrences à l'étude de certains systèmes de commande, Thèse de Docteur-ingénieur, Toulouse, (1969).
62. BERNUSSOU J., LIU HSU, MIRA C., Quelques exemples de solutions stochastiques bornées dans les récurrences autonomes du deuxième ordre. Actes du coloque international "Transformations ponctuelles et applications", Toulouse, (1973).
63. PICARD A., Leçons sur quelques équations fonctionnelles, Gauthier Villars, Paris, (1950).

64. ARNOLD V.I., AVEZ A., Problèmes ergodiques de la mécanique classique, Gauthiers Villars, Paris, (1967).
65. ABATUT J.L., DE COLIGNY M., Domaine d'attraction d'une singularité d'une récurrence non linéaire du 2ème ordre, RAIRO, n° 1, p. 77 - 80, (1972).
66. GUMOWSKI I., MIRA C., Détermination des courbes invariantes fermées d'une récurrence ou transformation ponctuelle non linéaire voisine d'une récurrence conservative linéaire, C.R.A.S. Paris, t. 276, Série A, (1973).
67. NEIMARK Yu.I., Izvestia vusov radiofisica, USSR, t. 1, n° 2, (1958), in russian.
68. GUMOWSKI I., MIRA C., Accumulations de bifurcations dans une récurrence, C.R.A.S. Paris, t. 281, Série A, (1975).
69. OSTER G., GUCKENHEIMER J., Bifurcation behaviour of population models ; in the Hopf bifurcation, J. Marsden and M. Mc Cracken eds, Springer Verlag, New York, (1976).
70. MINORSKI N., Non Linear oscillation, D. Van Nostrand Company inc., (1962).
71. MIRA C., Traversée d'un cas critique pour une récurrence du 2ème ordre sous l'effet d'une variation de paramètre, C.R.A.S., t. 268, p. 621 - 624, Série A, (1969).
72. NEIMARK Yu.I., Izvestia vusov radiofisica, USSR, t. 11, n° 11, (1968).
73. RUELLE D., TAKENS F., On the nature of turbulence. Com. Math. Phys. 20, p. 167 - 192, (1971).
74. PRAJOUX R., GIRAUD A., VALETTE R., A modelling technique using a recurrence for some control systems described by a piecewise-time-invariant-continuous state equation, Conference on "Information sciences and systems", John Hopkins University, Baltimore, 2.4 April, (1975).
75. ROZENVASSER, General sensitiviy equations of discontinuous systems, Automation and remote control, n° 3, (1967).
76. LEFSCHETZ S., Differential equations, geometric theory, Interscience, New York, (1957).
77. BIRKHOFF Garret, On product intergration, Journal Math. Phys., Vol. 16, p. 104 - 132, (1937).

AUTHOR INDEX

Numbers in parentheses are reference numbers, the others show the page where an author's work is referred to in the text

SUBJECT INDEX